TI-81 GRAPHING ACTIVITIES FOR APPLIED CALCULUS

TI-81 GRAPHING CALCULATOR ACTIVITIES FOR APPLIED CALCULUS

Wayne L. Miller
Donald Perry
Gloria A. Tveten

Lee College
Baytown, Texas

Brooks/Cole Publishing Company
Pacific Grove, California

Brooks/Cole Publishing Company
A Division of Wadsworth, Inc.

Printed in the United States of America

10 9 8 7 6 5 4 3 2 1

Library of Congress Cataloging-in-Publication Data
Miller, Wayne (Wayne L.)
    TI-81 graphing calculator activities for applied calculus /
Wayne L. Miller, Donald Perry, Gloria A. Tveten.
       p. cm. — (Graphing calculator series)
    ISBN 0-534-17462-0
    1. Calculus—Data processing.   2. TI-81 (Calculator)   I. Perry,
Donald.   II. Tveten, Gloria A., [date]–   .   III. Title.   IV. Series.
QA303.5.D37M55 1992
512'.0285'41—dc20

                                                92-15299
                                                  CIP

Sponsoring Editor: *Paula-Christy Heighton*
Marketing Representative: *Ragu Raghavan*
Editorial Assistant: *Carol Ann Benedict*
Production Coordinator: *Dorothy Bell*
Interior and Cover Design: *E. Kelly Shoemaker*
Printing and Binding: *Malloy Lithographing, Inc.*

*We dedicate this book to our colleagues
Bob Seale, Earl Hines, and Jean Norred.*

*We appreciate their help, support, and
understanding through the many years
we have worked together.*

*We also dedicate this book to the memory
of two special colleagues,*

Charles Bonner and Florence Neill.

*Their guidance and support when we came
to Lee College was invaluable.*

Preface

Modern technology has advanced to the stage where it should be used by mathematics students at all levels of study. The new calculators with graphing, programming, statistical, and matrix capabilities open avenues of exploration that have not been possible in the traditional classroom setting. The mathematics classroom can become a mathematics laboratory, where topics are explored and understanding is reinforced. Applications can become much more realistic. The study of traditional topics can be enhanced because many of the artificial boundaries that limited the teacher's and the student's capabilities have been removed.

The ability to see graphs instantly expands the study of the graphing of equations and inequalities. No longer do students or teachers have to laboriously plot points or memorize techniques for graphing certain families of functions or relations. Instead, the class can use the ability to graph functions with ease to investigate such topics as limits and continuity of functions. The relationships between functions and their derivatives with respect to maxima or minima, inflection points, concavity, and other topics can also be explored. The use of Taylor or Maclaurin series to approximate non-algebraic functions can more easily be understood with the help of the graph.

The study of programming will be a new topic for many mathematics classes, but it is one that needs to be explored. The calculator's programming capabilities illustrate how programming can be used as a tool to solve other problems in mathematics. For example, Newton's method for finding roots of a function can be easily programmed into the calculator with a minimum of programming instruction. Combined with programming, numerical integration using methods such as Simpson's rule can be an effective tool for approximating definite integrals. When programming is combined with graphing, the area between two curves can not only be found using numerical integration, it can also be illustrated with a graph. The ability of the calculator to approximate the derivative of a function at a point can be combined with graphing and programming to create the graph of the derivative without having to find the derivative algebraically.

The field of statistics has applications in many mathematics classes. Using the calculator to calculate factorials, permutations, and combinations can save valuable time. The study of one-variable statistics is enhanced with the ability to input data into the calculator and instantly have available the mean, the population and sample standard deviations, a histogram, or a frequency polygon. For two-variable statistics, the calculator has available four different regression equations and the correlation coefficient for each. Using programming and numerical integration, continuous probability density functions can be examined without having to integrate the functions.

The ability to use matrices to store data for later recall can be an added asset. As a program generates data, such as the value of a function as x approaches a limit, the data can be stored by the program. Topics that have had to be examined abstractly or only after laborious calculations can now be studied fully.

One other important aspect of the new calculators needs to be emphasized. Since all work is done using a viewing screen, which is somewhat like a small computer monitor screen, errors in calculation can be more easily located. Using the traditional scientific calculator, only the results of the keystrokes are seen. With graphing calculators, the operation or entry is seen on the screen and can be corrected, if necessary, before the operation is performed. Thus, the understanding of such topics as the order of operations is enhanced.

This textbook has been written to help students understand how to use the graphing calculator with a minimum of instruction or support from the teacher. The narrative in the text uses an informal, but informative, style of writing. Technical jargon is minimized. The graphing, editing, and advanced function keys are explained in detail. The advanced function keys give access to menus, and each item in these menus is discussed. As each key or menu item is discussed, examples are included, and the exact keystrokes necessary to execute the examples are given. The authors assume that the student is familiar with the traditional scientific calculator, and the only keys discussed are those that are new to the graphing calculator.

The material covered in the text is grouped into five chapters, each involving a single central theme. The first chapter covers the basic aspects of graphing. Chapter 2 discusses how to program the calculator. Chapter 3 covers the statistical capabilities of the calculator. Chapter 4 explains the use of matrices. Chapter 5 contains term projects.

Problem sets designed to help students better understand how to use the calculator are included in the first four chapters. Since the focus of this book is also directed toward the understanding of the mathematical principles involved, some of the problem sets instruct students to write a paragraph explaining the procedure used in solving the problem. Answers to selected exercises are given at the end of the book.

The authors hope that this text will prove helpful to both teachers and students. In today's world, one cannot function successfully without a good foundation in mathematics. If the use of this text and the graphing calculator helps the teacher to build a better foundation for his or her students, then the hope of the authors will be realized.

Wayne L. Miller
Donald Perry
Gloria A. Tveten

Lee College
Baytown, Texas

Contents

CORRELATION CHART
APPLIED CALCULUS – TI-81 CALCULATOR HANDBOOK
(Section numbers refer to the textbooks. Page numbers refer to the calculator handbook.)

TI-81 GRAPHING CALCULATOR ACTIVITIES	APPLIED CALCULUS, 2nd Ed. by Taylor and Gilligan	CALCULUS WITH APPLICATIONS, 2nd Ed. by Smith
Chapter 1 Graphing Capabilities	Sec. 1.3-1.4 can be done if students know how to graph equations pp. 5-13, 30, 34-37 Sec. 1.5 (Graphing) pp. 5-13 Sec. 2.1 (Graphing) pp. 5-13, 14-15 Sec. 2.2-2.4 pp. 16-20 Sec. 4.1-4.3 pp. 21-23 Sec. 4.5 pp. 5-13, 21-23 Sec. 5.1-5.2 (Graphing and Equations) pp. 5-7 Sec. 6.1 p. 23	Sec. 1.2-1.5 pp. 5-13, 14-15 Sec. 2.1-2.2 pp. 16-20 Sec. 4.1-4.2 pp. 21-23 Sec. 4.3 pp. 5-13, 21-23 Sec. 5.1-5.2 pp. 5-13 Sec. 5.4 pp. 5-13, 21-23 Sec. 6.1 p. 23 Sec. A.5-A.6 can be done if students know how to graph equations pp. 5-13, 30, 34-37
Chapter 2 Programming Capabilities	Sec. 2.5 pp. 55-56 Sec. 2.6 pp. 49-53 Sec. 6.4 pp. 58-60 Sec. 7.3-7.4 pp. 56-61	Sec. 2.4 pp. 49-53, 55-56 Sec. 6.4 pp. 58-60 Sec. 6.6 pp. 56-58 Sec. 7.4 pp. 60-61
Chapter 3 Statistical Capabilities	Sec. 7.4 pp. 82-84 Sec. 9.6 pp. 81-82	Sec. 7.5 pp. 82-84 Sec. 8.6 pp. 81-82

Chapter 1   GRAPHING CAPABILITIES

In applied calculus we discuss functions, the derivatives of functions, and the antiderivatives (integrals) of functions as well as the applications of each. We can use graphing to aid us in understanding these functions. We can also use graphs to discuss characteristics such as the limit, critical values, concavity, slope, domain and range, continuity, and maximum or minimum values of functions. Before we are ready to graph functions, however, we need to learn to use several keys on the calculator.

USING THE GRAPHING CALCULATOR

Turning the Calculator On and Off

To turn the calculator on, press the ON key. To turn the calculator off, press the 2nd key followed by the ON key. Pressing the 2nd key followed by another key, such as the ON key, activates the calculator instruction written in blue letters to the left and above the key pressed after the 2nd key. Thus, 2nd ON is interpreted by the calculator as OFF. The calculator also has a battery-saving feature that automatically turns the calculator off after about five minutes of no activity. If the calculator automatically turns itself off, it can be turned back on with the ON key and the work will still be on the viewing screen.

Adjusting the Viewing Screen

The viewing screen can be lightened or darkened by pressing the 2nd key and then holding down the $\triangle$ or $\triangledown$ key. The $\triangle$ key will darken the viewing screen and the $\triangledown$ key will make it lighter. If the screen must constantly be darkened, the batteries probably need to be replaced.

The MODE Key and the Cursor Keys

Press the MODE key. The MODE key determines how numbers and graphs will be displayed and calculated. We will be concerned primarily with how graphs are displayed. On the viewing screen, there should be eight lines of information with the current settings on each line highlighted. There should also be a blinking rectangle at the current setting for the first line. This blinking rectangle can be moved horizontally along a line using the ◁ and ▷ keys and vertically to another line using the △ and ▽ keys. These four keys are called the cursor keys. To change the setting on a particular line, move the blinking rectangle to the desired setting and then press the ENTER key. Now let's examine each line of information and determine the settings needed for our graphing.

The first line is the type of notation to be displayed. The **Norm** setting should be highlighted. Numbers will be displayed in their normal or usual way. The **Sci** setting displays numbers in scientific notation. In the **Eng** setting, the numbers may have one, two, or three digits to the left of the decimal point. The power-of-ten exponent is always a multiple of three.

The second line determines the number of decimal places to be displayed. The **Float** setting should be highlighted. The **Float** setting allows the number of decimal places to vary from number to number as needed. The other settings fix the number of decimal places to be displayed. For example, if the **4** setting is highlighted, then every number will be displayed with four decimal places. This can result in unnecessary zeros on a number.

The third line is the type of trigonometric angle measurement the calculator is to use and display. The **Deg** setting means that angles will be displayed in degrees. The **Rad** setting means that angles will be displayed in radian measure. Highlight the **Rad** setting because in calculus we assume that angles are measured in radians.

The fourth line tells the calculator the type of graphing we will be doing.

The **Function** setting, which should be highlighted, tells the calculator that we will be entering our equations as y in terms of x, as in $y = x^2 - 3x$. The **Param** setting is used for parametric equations. In parametric equations, both x and y are expressed in terms of a third variable t, called a parameter. To enter an equation such as $y = x^2 - 3x$, we will use the $\boxed{\text{X}|\text{T}}$ key. With the **Function** setting highlighted, the calculator will put an X on the viewing screen and not a T.

The calculator graphs a function by calculating points, values for x and y, at preselected intervals on the x-axis. In the fifth line, if the **Connected** setting is highlighted, the calculator will draw a line between adjacent points. If the **Dot** setting is highlighted, the calculator will plot the dots but will not connect them. Highlight the **Connected** setting.

The calculator can graph up to four different equations at the same time. On the sixth line, if the **Sequence** setting is highlighted, the calculator will draw the graphs sequentially; that is, it will graph the second one after it finishes with the first, and so forth. If the **Simul** setting is highlighted, the calculator will draw all of the graphs at the same time. Let's highlight the **Sequence** setting because multiple graphs are easier to recognize and interpret if each one is graphed individually.

When we draw a graph, we will tell the calculator how to scale the x- and y-axes — that is, whether we want each mark on a particular axis to represent 1 unit, 5 units, 10 units, and so on. The seventh line has **Grid Off** and **Grid On** settings. The **Grid On** setting means that grid points are displayed on the graph. These grid points correspond to the intersection points of the lines on a sheet of graph paper. The grid points can be used to estimate the value of points on the graph. The **Grid Off** setting means that the grid points are not displayed. Highlight the **Grid Off** setting. We can more accurately estimate the value of points using the $\boxed{\text{Zoom}}$ and $\boxed{\text{Trace}}$ keys, which we will discuss later.

The eighth line determines whether we will be graphing in the rectangular or the polar coordinate system. Highlight the **Rect** setting. The **Polar** setting is used for graphing certain types of trigonometric equations.

```
Norm Sci Eng
Float 0123456789
Rad Deg
Function Param
Connected Dot
Sequence Simul
Grid Off Grid On
Rect Polar
```

To summarize, we should have the **Norm**, **Float**, **Rad**, **Function**, **Connected**, **Sequence**, **Grid Off**, and **Rect** settings highlighted. Remember, to change the highlighted setting on any line, use the cursor keys to move the blinking rectangle to the new setting on the line and press the ENTER key to store that change in the calculator's memory. After highlighting the correct settings, exit the Mode screen by pressing the CLEAR key. This tells the calculator that we are finished working in the Mode screen. It can be done from any line on the Mode screen. The calculator then returns to the Home screen.

The RANGE Key

The RANGE key is used to determine the minimum and maximum values for the x- and y-coordinates and the scale that will be displayed on the viewing screen. Press the RANGE key. On the viewing screen we will see the word RANGE followed by seven lines of variables, with each variable followed by an "=" sign and a number. The numbers are the current values for the variables. We can change the value for any variable by moving up and down the list using the △ and ▽ keys, entering the desired number, and then pressing the ENTER key. The ◁ and ▷ keys are used to change digits within a number. We can leave the Range screen by pressing the 2nd and CLEAR keys.

Now let's set the values for each of the variables. Move the blinking rectangle so that it is on the first digit or sign of the **Xmin** variable. Let's set **Xmin** $= -5$. To put a "negative" sign on a number we use the (−) key and not the − key in the last column. The keystrokes are (−), 5, and ENTER. The blinking rectangle should now be on the **Xmax** variable. Set **Xmax** $= 5$ and **Xscl** $= 1$. The keystrokes are 5, ENTER, 1, and ENTER. The blinking rectangle should now be on the **Ymin** variable. Before entering the values for the remaining variables, let's review what we have done. We have defined the minimum and maximum values for the variable x to be shown on the viewing screen as -5 and 5. We have set the scale on the x-axis as 1, so each mark on the x-axis will represent 1 unit. Now let's set **Ymin** $= -5$, **Ymax** $= 5$, and **Yscl** $= 1$. Again, the keystrokes are (−), 5, ENTER, 5, ENTER, 1, and $\boxed{\textbf{ENTER}}$. The minimum values, maximum values, and the scale for x and y do not necessarily have to be the same. Some graphs might warrant values such as $\textbf{Xmin} = -5$, $\textbf{Xmax} = 15$, $\textbf{Xscl} = 1$, $\textbf{Ymin} = -20$, $\textbf{Ymax} = 70$, and $\textbf{Yscl} = 10$. The rule of thumb for the minimum value for the scales is (Xmax $-$ Xmin)/20 or (Ymax $-$ Ymin)/20. Using a smaller value will generally cause the marks on the axes to be so close together that they cannot be easily read. Finally, the blinking rectangle should now be on the value for the $\textbf{Xres}$ variable. It can be set for any value from 1 to 8. Generally, the setting $\textbf{Xres} = 1$ is used because the x-axis is divided into 96 dots and if $\textbf{Xres} = 1$, then the calculator plots all 96 dots. If $\textbf{Xres} = 2$, then the calculator plots every other dot, or 48 in all, and the graph is not as smooth. Setting the $\textbf{Xres}$ at any value larger than 1 speeds up the graphing process because the calculator has fewer points to calculate and plot, but the graph is usually distorted.

```
RANGE
Xmin=-5
Xmax=5
Xscl=1
Ymin=-5
Ymax=5
Yscl=1
Xres=1
```

Problem Set 1 ——

1. Describe what you think would happen if you set $\textbf{Xmin} = 5$ and $\textbf{Xmax} = -3$. Now change the Range variables to these values, press the $\boxed{\textbf{GRAPH}}$ key, and see what happens.

2. Reset the Range values for $\textbf{Xmin}$ and $\textbf{Xmax}$ and describe what you think would happen if you set $\textbf{Ymin} = 5$ and $\textbf{Ymax} = 5$. Now change the Range variables to these values, press the $\boxed{\textbf{GRAPH}}$ key, and see what happens.

Before proceeding, reset the Range variables to $\textbf{Xmin} = -5$, $\textbf{Xmax} = 5$, $\textbf{Ymin} = -5$, and $\textbf{Ymax} = 5$.

The $\boxed{\textbf{Y=}}$ Key

Before discussing the remaining keys, we need to learn how to graph an equation. Start by pressing the $\boxed{\text{Y=}}$ key. On the viewing screen we will see four colons followed by $Y_1 =$, $Y_2 =$, $Y_3 =$, and $Y_4 =$. We can enter up to four equations and graph all of them at the same time. If there is anything on any of these lines after the "=" sign, it can be erased by pressing the $\boxed{\text{CLEAR}}$ key followed by the $\boxed{\text{ENTER}}$ key. Return to the first line using the $\boxed{\triangle}$ key. We will enter the equation $y = x^2 - 3x$ and let the calculator draw its graph. Position the blinking rectangle after the "=" sign on the first line. Now press the keys $\boxed{\text{X|T}}$, $\boxed{x^2}$, $\boxed{-}$, $\boxed{3}$, $\boxed{\text{X|T}}$, and $\boxed{\text{ENTER}}$ in that order. The blinking rectangle should now be after the "=" sign on the second line. The first line should read $Y_1 = X^2 - 3X$ with the "=" sign highlighted.

```
:Y₁▉X²−3X
:Y₂=
:Y₃=
:Y₄=
```

If we make a mistake we can use the cursor keys — $\boxed{\triangle}$, $\boxed{\triangledown}$, $\boxed{\triangleleft}$, or $\boxed{\triangleright}$ — to position the blinking rectangle over the error and type the correct key. After the correction, press $\boxed{\text{ENTER}}$ to store the equation in the calculator's memory. The $\boxed{\text{INS}}$ and $\boxed{\text{DEL}}$ keys can also be used to make corrections. If we had accidentally left out the 3 in 3X, we could have used the cursor keys to position the blinking rectangle on top of the X, pressed the $\boxed{\text{INS}}$ key, pressed the $\boxed{3}$ key, and the calculator would have inserted a 3 before the X. Then we could have pressed the $\boxed{\text{ENTER}}$ key to store the corrected equation in the calculator's memory. Likewise, if we had accidentally typed 23X, we could have positioned the blinking rectangle over the 2 using the cursor keys, pressed the $\boxed{\text{DEL}}$ key, and the calculator would have deleted the 2. Then pressing the $\boxed{\text{ENTER}}$ key would have stored the corrected equation in the calculator's memory.

Usually equations are entered into the calculator as they are written on paper. The calculator knows that 3X means "3 times X" and we do not have to enter the "times" sign which looks like *. We have to use it only when we are multiplying two numbers. Likewise, when we press the $\boxed{x^2}$

key, the calculator knows that we are squaring the preceding number or variable and so it puts a "square" on only that number or variable. All of the black calculation keys work the same way. For example, if we press the $\boxed{x^{-1}}$ key, the calculator will put an exponent of -1 on the variable or number immediately preceding. Except in a very few special cases, we cannot enter a "Y" to the right of the "=" sign, so all equations must be solved for y in terms of x. We can enter the X by pressing the $\boxed{\text{ALPHA}}$ key followed by the $\boxed{\text{STO}}$ key, but that is not as fast as using the $\boxed{\text{X|T}}$ key. The calculator uses the standard order of operation rules, so we must be careful to put parentheses in the equations as needed to ensure that we get the correct result. This is especially true for rational equations. If we wish to graph $y = \frac{x-3}{x+2}$, we must enter it in the calculator as $Y_1 = (X-3)/(X+2)$ and not as $Y_1 = X - 3/X + 2$.

Problem Set 2

For each of the following problems, write a paragraph describing the process used to input the first equation. Then describe the process used to change the first equation to the second equation with the $\boxed{\text{INS}}$ and $\boxed{\text{DEL}}$ keys.

1. Input: $\qquad y = x^2 + 7x - 5 + 4/x + 2$
   Change to: $\quad y = x^2 + 7(x-5) + 4/(x-2)$
2. Input: $\qquad y = x^3 + 7x^2 - 5x + 2$
   Change to: $\quad y = x^4 + 7x^2 + 5x - 2$
3. Input: $\qquad y = 2x + 3x + 8 + 3x + 7x + 3$
   Change to: $\quad y = \dfrac{2x+3}{x+8} + \dfrac{3x+7}{x+3}$

The $\boxed{\text{GRAPH}}$ Key

To graph $y = x^2 - 3x$, press the $\boxed{\text{GRAPH}}$ key. On the viewing screen, the calculator draws a set of axes whose minimum and maximum values and scale match the values we set using the $\boxed{\text{RANGE}}$ key. The calculator draws the graph from left to right. Note that the graph is not actually a smooth curve but rather a series of short lines that are vertical or horizontal depending on their position on the graph. The graph is a parabola and the x-intercepts seem to be at the origin, the point $(0,0)$, and 3 units to the right of the origin, the point $(3,0)$. There are three methods of finding the vertex of the graph: using the cursor keys, using the TRACE key, and using the ZOOM key.

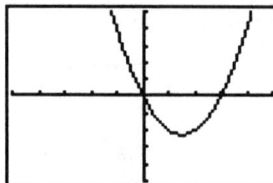

Using the Cursor Keys to Locate Points

The first method is to use the cursor keys. If we press the ▷ key twice, we will see blinking crosshairs that appear to be on the x-axis. On the lower part of the screen we will see $x = .26315789$ and $y = .07936508$.

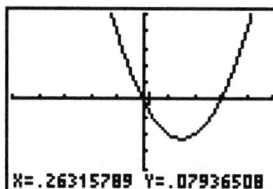

These are the x- and y-coordinates of the point that is blinking. From the y value of the point we know that it is slightly above the x-axis. Now press the ▷ key again. The x changes to $x = .36842105$, but the y stays the same.

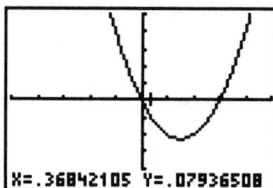

The difference between the two x values is 0.10526316. If we press the ▷ key again, the x changes to $x = .47368421$ and again y stays the same. The difference between this value for x and the one immediately to the left is again 0.10526316. The number is calculated using the formula $(\text{Xmax} - \text{Xmin})/95$. Earlier, when we discussed the setting for the **Xres**

variable, we said that the viewing screen was divided horizontally into 96 dots. The number 0.10526316 is the horizontal width of each dot. From the point $x = .47368421$ and $y = .07936508$, press the $\boxed{\triangle}$ key. This time the x stays the same and the y changes to $y = .23809524$. The difference between these two y values is 0.15873016, which is the vertical width of each dot. The formula $(\text{Ymax} - \text{Ymin})/63$ calculates this value because the viewing screen is divided vertically into 64 dots. Thus the viewing screen is divided into 6144 points (96×64). When we set the minimum and maximum values for x and y using the $\boxed{\textbf{RANGE}}$ key, we determine the coordinates of each of the 6144 points. To graph an equation, the calculator substitutes each of the values it has for x into the equation and calculates a corresponding value for y. It then "plots" or "lights up" the point whose y value most closely matches the value for y it has calculated. This means that we may not be able to locate the coordinates of a particular point on the graph exactly, because the coordinates of that point may not be one of the 6144 points on the viewing screen. To illustrate this, let's use the cursor keys to move the blinking crosshairs to the vertex of the curve. At the point $(1.5263158, -2.142857)$, the crosshairs seem to be at the vertex. The vertex is actually at the point $(1.5, -2.25)$, but the graph does not include that point as one of the 6144 points. Sometimes the graph will include a particular point and sometimes we can only estimate the point. Later we will discuss a method for improving an estimate.

Problem Set 3

1. Set **Xscl** $= 1$. Choose values for **Xmin** and **Xmax** so that if we use the cursor keys to move the blinking crosshairs along the x-axis to a tic mark, the value for x displayed on the viewing screen is an integer.
2. Set **Yscl** $= 1$. Choose values for **Ymin** and **Ymax** such that the values for y displayed on the screen at the tic marks are integers.

Before proceeding, reset the Range variables to **Xmin** $= -5$, **Xmax** $= 5$, **Ymin** $= -5$, and **Ymax** $= 5$.

Using the ⏹TRACE⏹ Key to Locate Points

Press the ⏹TRACE⏹ key and a little blinking rectangle near the intersection of the coordinate axes will appear on the screen. We can move this rectangle along the curve using the ⏹◁⏹ and ⏹▷⏹ keys. Use the ⏹▷⏹ key to move the rectangle until it appears to be at the vertex. The coordinates of the point at the bottom of the viewing screen should now be $x = 1.5263158$ and $y = -2.249307$.

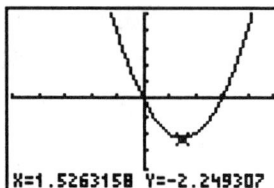

X=1.5263158 Y=-2.249307

Note that the x-coordinate is the same as that found when we used the cursor keys to locate the vertex, but the y-coordinate is different. If we use the ⏹TRACE⏹ key to locate a point on the graph, then each time we change the x-coordinate, the calculator calculates the y-coordinate for that value of x. In this case, the calculator substitutes $x = 1.5263158$ into $y = x^2 - 3x$ and gets $y = -2.249307$. The estimate using the ⏹TRACE⏹ key is usually more accurate than the one found using the cursor keys.

If we are graphing more than one equation, we can use the ⏹△⏹ and ⏹▽⏹ keys to move the blinking rectangle from one equation to another. If the y value of an equation is undefined at a particular x value, then the y value on the viewing screen is left blank. If we are tracing a curve and the y value is smaller than the **Ymin** we set or larger than the **Ymax** we set, the blinking rectangle will disappear, but the coordinate values will continue to be displayed at the bottom of the viewing screen. If the x value gets outside the range we set for x, then the calculator automatically adjusts the viewing screen and updates the values for **Xmin** and **Xmax**.

Using the ⏹ZOOM⏹ Key to Locate Points

The ⏹ZOOM⏹ key, which can also be used to locate points on a graph, has several features that are not available using the cursor keys or the TRACE key. Press the ZOOM key. On the viewing screen, the word ZOOM is followed by a menu of instructions that can be accessed either by using the △ and ▽ keys followed by the ENTER key or by pressing the number key that corresponds to the number of the instruction. The menu has eight items, but since the viewing screen can accommodate only seven lines, the "↓" on the seventh instruction tells us that there is at least one more item. To see it, scroll down the screen using the ▽ key. We will discuss only the first seven instructions and will usually use only the first three. The items in the menu are

1 : Box
2 : Zoom In
3 : Zoom Out
4 : Set Factors
5 : Square
6 : Standard
7 : Trig
8 : Integer

The first instruction is **Box**. Choosing it allows us to draw a box around a particular portion of the curve. The calculator will then change the Range values to match the box and redraw the curve using these values. Thus the portion of the curve in the box is enlarged. Now press the ① key. We will see a set of blinking crosshairs like those seen when we were using the cursor keys. Now let's create a box around the vertex. Use the cursor keys to move the crosshairs to a point a little above and to the left of the vertex and press the ENTER key. Note that the blinking crosshairs change to a blinking rectangle. Now move the blinking rectangle to a point a little below and to the right of the vertex. As we do this, the calculator draws a box around the vertex. Now press the ENTER key again. The calculator redraws and enlarges the portion of the curve in the box, so that it fills the viewing screen.

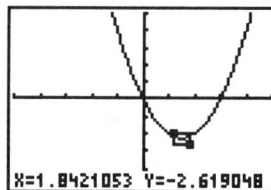

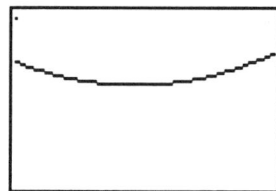

If we press the RANGE key, we will see that the minimum and maximum values for x and y have been adjusted to match the boundaries of the box. If we press the GRAPH key again, the graph of the equation will reappear. We can then use the TRACE key to get a better approximation for the vertex, or the **Box** instruction to further enlarge the graph. Before going to the next instruction, press the RANGE key and reset **Xmin** $= -5$, **Xmax** $= 5$, **Ymin** $= -5$, and **Ymax** $= 5$. Then press the GRAPH key to redraw the graph.

Press the ZOOM key again. Now let's examine the **Zoom In** instruction. This allows us to zoom in on a particular portion of the curve without having to draw a box. The calculator automatically changes the viewing screen by a set factor. Press the **2** key and move the blinking crosshairs to the vertex and press the ENTER key. The calculator redraws the curve with this point at the center of the viewing screen. It also automatically adjusts the Range values to match the limits of the viewing screen. We can now move the blinking crosshairs to a better approximation of the vertex. Press the ENTER key again and the calculator will further magnify that portion of the graph. We can also use the TRACE key to estimate the vertex.

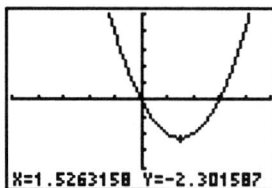

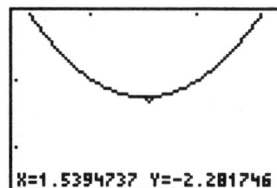

Press the ZOOM key again. The **Zoom Out** instruction acts exactly like the **Zoom In** instruction except that it displays a greater portion of the graph by a set factor rather than magnifying a portion of the graph. It is used primarily when the part of the graph to be examined is not on the viewing screen.

The portion of the graph seen on the viewing screen can be changed by adjusting the Range values and redrawing the graph using the GRAPH key instead of using the **Zoom In** or **Zoom Out** instructions. If the minimum and maximum values are brought closer together, the graph is reduced; if they are spread apart, the graph is enlarged. If we choose to do either of these, we may want to change the scale values.

The **Set Factors** instruction changes the ratio by which the graph is magnified or reduced when the **Zoom In** or **Zoom Out** instructions are used. Zoom factors are positive numbers equal to or greater than 1. The factors for x and y do not have to be the same. They are both currently set at 4, which means that the graph is magnified to four times its size or reduced to one-fourth its size.

The **Square** instruction adjusts the width of the dots and updates the Range variables so that the x and y are equal. Since the viewing screen is not square, some graphs, such as a circle, appear distorted. The **Square** instruction corrects that. Since this will not be a problem with any of our graphs, we will not use this instruction.

The **Standard** instruction sets the minimum and maximum values for x and y to -10 and 10 and sets the scale values to 1. These are called the standard values for the viewing screen. We will use this instruction from time to time.

The **Trig** instruction resets the Range variables to preset values appropriate for trigonometric functions. Since we have set the trigonometric angle measurement for radians, the calculator sets **Xmin** = -6.28318531 or -2π, **Xmax** = 6.283185307 or 2π, **Xscl** = 1.570796327 or $\pi/2$, **Ymin** = -3, **Ymax** = 3, and **Yscl** = 0.25.

Problem Set 4 ————————————————————————————

For each of the following problems, write a paragraph explaining how to graph the equation. Determine whether the entire graph of the equation can be seen on the viewing screen. If it cannot, determine how to change the screen so that it can be seen. Locate the x- and y-intercepts and describe how the zoom, box, and trace functions helped locate them.

1. $y = x^3 + 7x^2 - 5x + 2$

2. $y = \dfrac{2x + 3}{x + 8} + \dfrac{3x + 7}{x + 3}$

3. $y = \pm \sqrt{1000 - x^2}$

Before proceeding, reset the Range variables to **Xmin** $= -5$, **Xmax** $= 5$, **Ymin** $= -5$, and **Ymax** $= 5$.

The TEST Key

Press the 2nd key followed by the MATH key. On the viewing screen we will see the word TEST followed by a menu of six relational operators. These are $=$, $\neq$, $>$, $\geq$, $<$, and $\leq$.

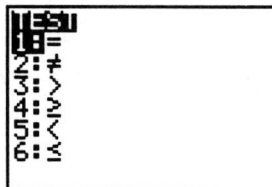

One use for these operators is to define the domain for certain types of functions we are graphing. We can select an operator from the list by pressing the number key associated with the operator or by using the ∇ key to scroll down the list and pressing the ENTER key when the number corresponding to the correct operator is highlighted. The selected operator is copied to the current cursor location in the expression we are writing. We can leave the TEST menu without making a selection by pressing the CLEAR key.

EXAMPLES OF GRAPHING APPLICATIONS

Graphing Functions with Restricted Domains

Suppose we wish to graph the function $y = 4x - x^2$ but wish the domain to be $0 \leq x \leq 4$. Press the Y= key. Clear all functions from the viewing screen. The keystrokes to enter the equation and the restricted domain are (, 4, X|T, −, X|T, x^2,), ((, 0, 2nd, MATH, 6, X|T,), ((, X|T, 2nd, MATH, 6, 4,)), and ENTER. On the viewing screen we should have $Y_1 = (4X - X^2)(0 \leq X)(X \leq 4)$. If the function to be graphed includes a $+$ or $-$, the function must be placed in a set of parentheses. If the set of parentheses is left out, the calculator assumes that the restriction on the domain applies only to that part of the function immediately preceding the restriction. The calculator cannot interpret the restriction $0 \leq x \leq 4$ stated as a double inequality. It must be divided into two separate inequalities, as indicated above. A single inequality such as $0 \leq x$ can be stated $x \geq 0$ if we wish to do so. Pressing the GRAPH key, we see that the graph is a parabola, but only that portion of the graph in the first quadrant is graphed. The restricted domain of $0 \leq x \leq 4$ eliminates that portion of the graph where $y < 0$.

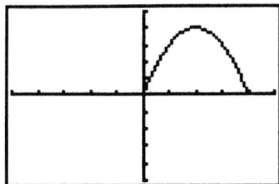

Restricted domains are also useful in graphing piecewise functions. A piecewise function has a graph that takes different shapes over different parts of its domain. The graph often includes a break or sharp turn. Let's graph the function

$$y = \begin{cases} x^2 - 3x & \text{for } x \leq 2 \\ 3 - x & \text{for } x > 2 \end{cases}$$

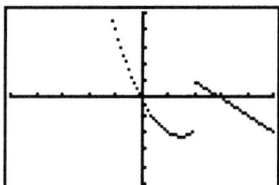

When graphing piecewise functions, we first press the MODE key and change the **Connected** setting to the **Dot** setting. In doing this, we prevent the calculator from drawing lines across any breaks that might occur in the graph. Press the Y= key. Clear all functions from the viewing screen. Enter the following equation and graph it. Each part of the function is followed by its domain, and the calculator uses the "$+$" sign to connect the parts of the function. The equation is

$$Y_1 = (X^2 - 3X)(X \leq 2) + (3 - X)(X > 2)$$

Problem Set 5

For each of the following functions, write a paragraph giving each keystroke used to input the equations and the restricted domains. Describe the screen after the final keystroke, and describe the graph of the function.

1. $$y = \begin{cases} x^2 + x^3 & \text{for} & 0 \le x \le 1 \\ x^2 & \text{for} & 1 < x \le 4 \\ x^{1/3} & \text{for} & 4 < x \le 8 \end{cases}$$

2. $$y = \begin{cases} \cos x & \text{for} & -\frac{\pi}{2} \le x \le \frac{\pi}{2} \\ \cos 2x & \text{for} & \frac{\pi}{2} < x \le \frac{3\pi}{2} \\ \sin x \cos x & \text{for} & \frac{3\pi}{2} < x \le 2\pi \end{cases}$$

3. $$y = \begin{cases} x & \text{for} & 0 \le x \le \frac{1}{8} \\ \frac{1}{8}\left(\frac{1-x}{7/8}\right) & \text{for} & \frac{1}{8} < x \le 1 \end{cases}$$

Before proceeding, press the $\boxed{\textbf{MODE}}$ key and change the **Dot** setting to the **Connected** setting.

Exploring Limits and Continuity of Functions Using Graphing

Let's graph the equation $y = x^2 - x - 1$. Press the $\boxed{\textbf{TRACE}}$ key and use $\boxed{\triangleright}$ key to move the blinking rectangle to the point on the curve where $x = 1.4210526$. If we continue to press the $\boxed{\triangleright}$ key five more times, we will see the values for x and y, rounded to three decimal places, change as follows

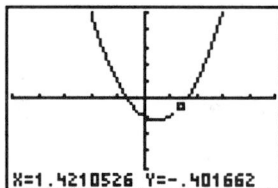

X=1.4210526 Y=-.401662

x	y
1.526	$-.197$
1.632	.030
1.737	.280
1.842	.551
1.947	.845

Continue to press the $\boxed{\triangleright}$ key until $x = 2.5789474$. Pressing the $\boxed{\triangleleft}$ key five times, we see the values for x and y, rounded to three decimal places, change as follows

x	y
2.474	2.645
2.368	2.241
2.263	1.859
2.158	1.499
2.053	1.161

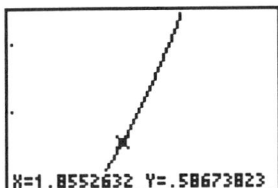

With the blinking rectangle at the last point, press the $\boxed{\text{ZOOM}}$, $\boxed{2}$, and $\boxed{\text{ENTER}}$ keys. The calculator redraws the graph. Press the $\boxed{\text{TRACE}}$ key again and use the $\boxed{◁}$ key to move the blinking rectangle to the point where $x = 1.8552632$. Pressing the $\boxed{▷}$ key five times, we obtain the following table of values

x	y
1.882	.659
1.908	.732
1.934	.807
1.961	.883
1.987	.961

Using the $\boxed{▷}$ key to move the blinking rectangle to $x = 2.1447368$ and pressing the $\boxed{◁}$ key five times, we get the following table of values

x	y
2.118	1.369
2.092	1.285
2.066	1.202
2.039	1.120
2.013	1.040

Examining the first and third tables, we see that as x approaches 2 from the left, y approaches 1. The second and fourth tables indicate the same result; as x approaches 2 from the right, y approaches 1. When the value of a function, y, approaches some number L as x approaches some number a from both sides of the number a, we say that "the limit of $f(x)$ as x approaches a is L" and we write: $\lim_{x \to a} f(x) = L$. In this case, we would write: $\lim_{x \to 2}(x^2 - x - 1) = 1$.

If we wanted further evidence that 1 is the limit of the function as x approaches 2, we could press the $\boxed{\text{ZOOM}}$, $\boxed{2}$, and $\boxed{\text{ENTER}}$ keys again and use the $\boxed{\text{TRACE}}$, $\boxed{◁}$, and $\boxed{▷}$ keys to generate the following two tables

x	y		x	y
1.970	.912		2.030	1.090
1.977	.931		2.023	1.070
1.984	.951		2.016	1.050
1.990	.970		2.010	1.030
1.997	.990		2.003	1.010

If we wish to discuss the limit of a function as x approaches some number *a* from only one side of the number, then we write $\lim_{x \to a^-} f(x) = L$ to discuss the limit as x approaches a from the left and $\lim_{x \to a^+} f(x) = L$ to discuss the limit as x approaches a from the right. In this case we would have $\lim_{x \to 2^-} (x^2 - x - 1) = 1$ and $\lim_{x \to 2^+} (x^2 - x - 1) = 1$. For a function to have a limit as x approaches some value a, the limit from both sides must exist and must be the same number.

As we discussed the limit of this function as x approaches 2, you might have observed that we would have gotten the same result if we had substituted 2 into the function and calculated the value of y. For many functions it is true that $\lim_{x \to a} f(x) = f(a)$; but it is not always true, as the next example illustrates.

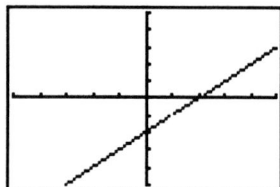

Reset the Range variables to **Xmin** $= -5$, **Xmax** $= 5$, **Ymin** $= -5$, and **Ymax** $= 5$. Enter $y = \dfrac{x^2 - 3x + 2}{x - 1}$ into the calculator and press the ⌐GRAPH⌐ key. You might be surprised to see that the graph is a straight line. If you examine the graph carefully, you will see that there is a "dot" missing from the line at the point $(1, -1)$. You can verify this by pressing the ⌐TRACE⌐ key and moving the blinking rectangle to the point where $x = 1$. There is no value for y on the screen because the function is undefined when $x = 1$. If we substitute 1 into the function for x, we get $\frac{0}{0}$. But if we move the blinking rectangle to the left and right of $x = 1$, we obtain the following values for x and y

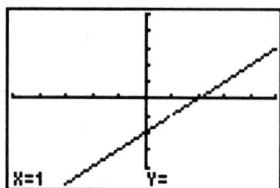

x	y
.474	-1.526
.579	-1.421
.684	-1.316
.789	-1.211
.895	-1.105

x	y
1.526	$-.474$
1.421	$-.579$
1.316	$-.684$
1.211	$-.789$
1.105	$-.895$

The limit of the function as x approaches 1 seems to be -1. We could use the ⌐ZOOM⌐ key with the **Zoom In** instruction to get values for x closer to 1, but they would only reinforce our assumption that the $\lim_{x \to 1} \dfrac{x^2 - 3x + 2}{x - 1} = -1$. In this case, although the function does not exist at a particular point, the limit does exist at that point.

It is also possible for a function to be defined at a particular point and not have a limit at that point. To illustrate this, let's go back to the piecewise function graphed earlier. Remember to change the **Connected** setting on the MODE key to the **Dot** setting. The function was

$$y = \begin{cases} x^2 - 3x & \text{for } x \le 2 \\ 3 - x & \text{for } x > 2 \end{cases}$$

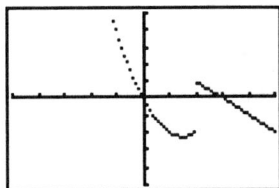

The function is defined for $x = 2$. Substituting 2 into the first equation, we have $y = -2$. But if we use the TRACE key and examine some values as x approaches 2 from both sides, we have

x	y		x	y
1.526	-2.249		2.474	.526
1.632	-2.233		2.368	.632
1.737	-2.194		2.263	.737
1.842	-2.133		2.158	.842
1.947	-2.050		2.053	.947

The limit as x approaches 2 from the left seems to be -2. The limit as x approaches 2 from the right seems to be 1. As we said earlier, if these two numbers are not the same, the function does not have a limit.

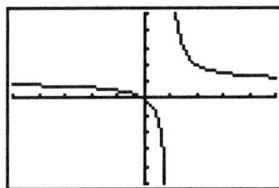

Press the MODE key and change the **Dot** setting back to **Connected**. Graph $y = \frac{x}{x-1}$. Press the TRACE key and move the blinking rectangle to the right. As x approaches 1 from the left, y becomes smaller. When $x = .89473684$, $y = -8.5$. The function does not approach a finite number, so it does not have a limit from the left. At $x = 1$, the function is not defined. As x approaches 1 from the right, y becomes larger. When $x = 1.1052632$, $y = 10.5$. Since it does not approach a finite number, it does not have a limit from the right. This is an example of a function that is not defined at a particular point and does not have a limit as x approaches that point. Since the function is not defined at a particular point, that point is not in the domain of the function, and the graph does not exist at that point. The vertical line through the point is called a vertical asymptote of the function. This function has a vertical asymptote at $x = 1$.

We can also use this graph to discuss two other types of limits. As we continue to move the blinking rectangle to the right, we can make two observations. First, as the blinking rectangle moves to the right edge of the viewing screen, the Range variables for **Xmin** and **Xmax** are adjusted so that the blinking rectangle remains on the viewing screen. Second, as x becomes larger, y approaches 1. Thus, $\lim_{x \to \infty} \frac{x}{x-1} = 1$. If we move the blinking rectangle toward the left edge of the viewing screen, the Range variables are readjusted. As x becomes a large negative number, y approaches 1. Thus, $\lim_{x \to -\infty} \frac{x}{x-1} = 1$. When the function approaches a finite value as x becomes a large positive or a large negative number, the horizontal line through this y value is called the horizontal asymptote of the function. This function has the horizontal asymptote $y = 1$.

We can also use these four functions to discuss the continuity of a function. The first function, $y = x^2 - x - 1$, is a continuous function. At each point a in the domain of the function, $\lim_{x \to a} f(x) = f(a)$. The last three have points of discontinuity. The second function, $y = \frac{x^2 - 3x + 2}{x - 1}$, has a point of discontinuity at $x = 1$. There is a "hole" in the graph at that point. The third function, the piecewise function, has a point of discontinuity at $x = 2$. The graph takes a "jump" at that point. The last function, $y = \frac{x}{x-1}$, has a point of discontinuity at $x = 1$. The graph of the function does not exist at that point.

Problem Set 6 ───

1. Write a plan to determine the limit of a function as x approaches a given value from the left and from the right.

2. Write a plan to determine the vertical and/or horizontal asymptotes of a function.

3. Write a plan to determine if a function is continuous or discontinuous over a given interval.

   Use your plans to work the following problems.

4. Find $\lim_{x \to 1} \frac{1 - x}{1 - \sqrt{x}}$

5. Find the vertical and horizontal asymptotes for $y = \frac{3x^3 - 2x - 4}{2x^3 - 3x^2 + 8}$, if they exist.

6. Find the discontinuities of the function $f(x) = \frac{8 - x^3}{x(x - 2)}$ from $(-\infty, \infty)$, if there are any.

Using Graphing to Explore Characteristics of Functions

Reset the Range variables to **Xmin** $= -4.5$, **Xmax** $= 5$, **Ymin** $= -10$, and **Ymax** $= 10$. Enter $y = x^3 - 2x^2 - 5x + 6$ on the first line of the Y = list. The keystrokes for entering x^3 are $\boxed{\text{X|T}}$, $\boxed{\text{MATH}}$, and $\boxed{3}$. Press the $\boxed{\text{TRACE}}$ key. The values chosen for **Xmin** and **Xmax** cause the values for x shown at the bottom of the viewing screen to change by 0.1 with each cursor keystroke. Using the $\boxed{\triangleleft}$ and $\boxed{\triangleright}$ keys, we find that the function has x-intercepts, or roots (points where the graph touches or crosses the x-axis), at $x = -2$, $x = 1$, and $x = 3$. The y-intercept occurs at the point $(0,6)$. We also find that the function is increasing for $x < -0.8$. For $-0.8 < x < 2.1$, the function is decreasing. For $x > 2.1$, the function is again increasing. A function is increasing on an interval if the slope of the tangent line to the function, called the slope of the function, is positive for each point in the interval. A function is decreasing on an interval if the slope of the tangent line to the function is negative for each point in the interval. The point where the function changes from an increasing function to a decreasing function is called a relative maximum point. This function has a relative maximum at about $x = -0.8$. The point where a function changes from a decreasing function to an increasing function is called a relative minimum point. This function has a relative minimum at about $x = 2.1$. At relative maximum and relative minimum points, the slope of the function is zero or undefined. Points where the slope of the function is zero or undefined are also called the critical values of the function. We could use the **Zoom In** instruction on the $\boxed{\text{ZOOM}}$ key to see if we could get a better approximation of the relative maximum and minimum points, but we will not do so at this time. Instead, we will use the derivative of the function, $\frac{dy}{dx} = 3x^2 - 4x - 5$, to help us examine these same ideas.

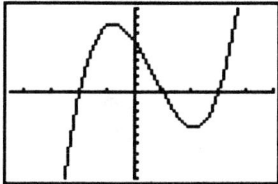

Enter $\text{Y}_2 = 3x^2 - 4x - 5$ on the second line of the Y = list and press the $\boxed{\text{GRAPH}}$ key. The graph is a parabola. Press the $\boxed{\text{TRACE}}$ key. Use the $\boxed{\triangledown}$ key to move the blinking rectangle to the parabola and use the $\boxed{\triangleleft}$ and $\boxed{\triangleright}$ keys to find the approximate values of the roots of the parabola. We find that the value for y is closest to zero when $x = -0.8$ and $x = 2.1$. It is no accident that these are the same points at which the relative

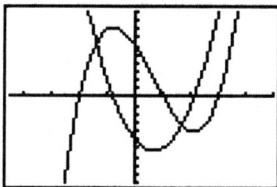

maximum and minimum points, respectively, occurred. The value of the derivative at any point is the value of the slope of the function at that point. If we examine the graph of the derivative, we find that for $x < -0.8$, the graph is above the x-axis, or $\frac{dy}{dx} > 0$. For these points, the function is increasing. For $-0.8 < x < 2.1$, the graph is below the x-axis, or $\frac{dy}{dx} < 0$. For these points, the function is decreasing. For $x > 2.1$, the graph is above the x-axis, or $\frac{dy}{dx} > 0$, and the function is increasing. Since the derivative of this function is a quadratic, we can set it equal to zero and solve for x using the quadratic formula. Correct to four decimal places, we have $x = -0.7863$ and $x = 2.1196$. These are the x-coordinates of the relative maximum and relative minimum points, respectively. We can substitute them into the function and obtain the y-coordinates. Now let's graph the second derivative of the function, $\frac{d^2y}{dx^2} = 6x - 4$, and discuss two more important ideas.

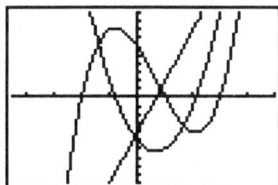

Enter Y3 = $6x - 4$ on the third line of the Y = list and press the GRAPH key. The graph is a line whose x-intercept is $x = \frac{2}{3}$, or $x = 0.6667$ correct to four decimal places. The second derivative measures the concavity of the function. In this case, when $x < 0.6667$ the second derivative is negative, or below the x-axis, and the function is concave downward. "Concave downward" means that the curve is shaped like a cup or bowl turned upside down. If the second derivative is negative, the point where the critical value occurs will be a relative maximum. For $x > 0.6667$, the second derivative is positive, or above the x-axis, and the function is concave upward. "Concave upward" means that the curve is shaped like a cup or bowl turned up. If the second derivative is positive, the point where the critical value occurs will be a relative minimum. The point where the second derivative is zero, $x = 0.6667$, is called an inflection point of the function. It is the point at which the function changes its concavity.

Let's examine a critical value where the first derivative does not exist by graphing a function, $y = (\sqrt[3]{x})^2$, and its derivative, $\frac{dy}{dx} = 1/\sqrt[3]{x}$.

Enter these into the calculator on the first two lines of the Y = list after changing the Range variables to **Ymin** = -3 and **Ymax** = 3. To enter

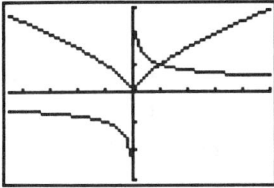

the cube root of the function press the MATH key followed by the 4 key. Press the GRAPH key. The function has a critical value which is a relative minimum point at $x = 0$ because the slope of the function changes from negative to positive. The graph of the function also makes a sharp turn at $x = 0$. If we use the TRACE key on the derivative, we will find that the derivative is undefined at $x = 0$. The derivative will be undefined at a point if the graph of the function makes a sharp turn at that point.

Using Graphing to Check Integrals of Functions

If we assign some arbitrary value for C, the constant of integration, we can then use the ideas developed in the previous section to determine if the indefinite integral of a function is correct. If $\int f(x)dx = F(x) + C$, then $\frac{dy}{dx}(F(x) + C) = f(x)$. We can graph the function and the integral and then determine if the criteria are satisfied for the function to be the derivative of the integral. For the function, let's use $f(x) = x^2 - 2x - 3$. The integral is $F(x) = \frac{1}{3}x^3 - x^2 - 3x + C$. Choose $C = 1$ (any value would do), set the Range variables to **Ymin** $= -10$ and **Ymax** $= 10$, enter the equations into the calculator with the function as Y₁, and press the GRAPH key. Using the TRACE key, we see that the slope of the integral is positive where the function is positive and negative where the function is negative. In addition, the points where the function is zero are the critical values of the integral. Since the function satisfies the criteria for the derivative of the integral, we have integrated the function correctly.

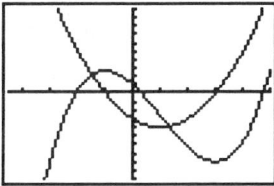

Using Algebraic Functions to Approximate Non-Algebraic Functions

Using a Taylor or Maclaurin series, we can develop algebraic functions which approximate non-algebraic functions. The series for a non-algebraic function is non-ending and the accuracy of the approximation is determined by the number of terms of the series included in the approximation. Graphing the non-algebraic function followed by functions which include an increasing number of terms can help us understand this concept. For example, the algebraic approximation of $y = e^x$ is

$$y = 1 + x + \frac{x^2}{2!} + \frac{x^3}{3!} + \frac{x^4}{4!} + \cdots$$

To approximate the function, let $Y_1 = e^x$, $Y_2 = 1 + x$, $Y_3 = 1 + x + \frac{x^2}{2!}$, and $Y_4 = 1 + x + \frac{x^2}{2!} + \frac{x^3}{3!}$. To enter the factorial sign, press the **MATH** key followed by the $\boxed{5}$ key. Change the Range variables to **Xmin** $= -3$, **Xmax** $= 3$, **Xscl** $= 1$, **Ymin** $= -5$, **Ymax** $= 15$, and **Yscl** $= 2$. As we can see from the graph, the first approximation is not very good, the second is better, and the third is even better. Increasing the number of terms in the algebraic approximation does improve the accuracy of the approximation.

Problem Set 7

1. Write a plan to determine the relative maxima, relative minima, concavity, and inflection points using the graphs of y, $\frac{dy}{dx}$, and $\frac{d^2y}{dx^2}$.
2. Write a plan that uses graphing to check the integration of functions.
3. Write a plan using a Taylor or Maclaurin series to approximate, to seven significant digits, the value of non-algebraic functions.

   Use your plans to work the following problems.

   For problems 4 and 5, find the relative maxima, relative minima, concavity, and inflection points.

4. $y = x^3 + 3x^2 + x - 6$
5. $y = x^4 + 2x^3 - 3x^2 - 4x + 4$
6. $\int \left(x + \frac{1}{x^2}\right) dx$. Show that your integration is correct.
7. Write an algebraic series to approximate $\cos x$.

USING THE GRAPHING CALCULATOR

The remaining keys discussed in this chapter will be used in examples in the following chapters. They are included here because of their relationship to graphing.

The MATH Key

```
MATH NUM HYP PRB
1:R▶P(
2:P▶R(
3: 3
4: 3√
5: !
6: °
7:↓r
```

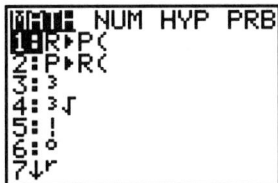

Press the MATH key. On the first line are the names of four menus that can be accessed by highlighting the name of the menu. These are MATH, NUM, HYP, and PRB. The PRB menu will be discussed in Chapter 3. The HYP menu is concerned with hyperbolic functions and the inverses of hyperbolic functions. These are not used in applied calculus and will not be discussed in this chapter.

The MATH Menu

The word MATH, which is an abbreviation for Mathematics, should be highlighted, so let's discuss it first. It is followed by an eight-item menu. We used some of the items in earlier examples, but we will explain them more fully now. The items are

$$1: R{\rightarrow}P($$
$$2: P{\rightarrow}R($$
$$3: {}^3$$
$$4: \sqrt[3]{}$$
$$5: !$$
$$6: {}^\circ$$
$$7: {}^r$$
$$8: \text{NDeriv}($$

The first item, **R→P(**, converts the rectangular coordinates of a point to the polar coordinates.

The second item, **P→R(**, converts the polar coordinates of a point to the rectangular coordinates. The form of the polar coordinates will depend on whether the MODE key is set for Radians or Degrees. Neither of these items will be needed for this course.

The third item finds the cube of the expression, variable, or quantity that precedes it. For example, if we wish to find the cube of 5, the keystrokes are 5, MATH, 3, and ENTER. The result is 125. We can achieve the same result using the ∧ key. The keystrokes would be 5, ∧, 3, and ENTER.

The fourth item finds the cube root of the expression, variable, or quantity that follows it. For example, if we wish to find the cube root of 64, the keystrokes are $\boxed{\text{MATH}}$, $\boxed{4}$, $\boxed{6}$, $\boxed{4}$, and $\boxed{\text{ENTER}}$. The result is 4. This result can also be achieved using the $\boxed{\wedge}$ key. The keystrokes would be $\boxed{6}$, $\boxed{4}$, $\boxed{\wedge}$, $\boxed{(}$, $\boxed{1}$, $\boxed{\div}$, $\boxed{3}$, $\boxed{)}$, and $\boxed{\text{ENTER}}$.

The fifth item, !, is called a factorial. The keystrokes for finding 5! are $\boxed{5}$, $\boxed{\text{MATH}}$, $\boxed{5}$, and $\boxed{\text{ENTER}}$ and the result is 120. Factorials are useful in calculating probabilities.

$$n! = n(n-1)(n-2)(n-3)\cdots 3\cdot 2\cdot 1. \text{ So } 5! = 5\cdot 4\cdot 3\cdot 2\cdot 1 = 120.$$

The sixth and seventh items on the list let us designate whether the expression, variable, or quantity that precedes it will be measured in degrees or radians. This designation overrides the current angle setting of the $\boxed{\text{MODE}}$ key for this expression, variable, or quantity only. Neither is needed for this course.

The eighth item, **NDeriv(**, is an abbreviation for Numerical Derivative. It calculates an approximation for the numerical value of the derivative of a function at a point x which has been stored in the calculator. The item requires two arguments, separated by a comma. The first argument is an expression in terms of x, such as $x^2 + x$, and the second argument is a small number, such as 0.01, called "delta x" or Δx. The item calculates the slope of the secant line through the points $(x - \Delta x, f(x - \Delta x))$ and $(x + \Delta x, f(x + \Delta x))$. This slope is an approximation for the numerical value of the derivative of the function. For example, let's store 2 for x in the calculator. The keystrokes are $\boxed{2}$, $\boxed{\text{STO}}$, $\boxed{\text{X|T}}$, and $\boxed{\text{ENTER}}$. Now let's approximate the numerical value of the derivative for $x^2 + x$ at $x = 2$ using 0.01 for Δx. The keystrokes are $\boxed{\text{MATH}}$, $\boxed{8}$, $\boxed{\text{X|T}}$, $\boxed{x^2}$, $\boxed{+}$, $\boxed{\text{X|T}}$, $\boxed{\text{ALPHA}}$, $\boxed{,}$, $\boxed{.}$, $\boxed{0}$, $\boxed{1}$, $\boxed{)}$, and $\boxed{\text{ENTER}}$. The result is 5. The accuracy of the result is determined by the size of the number we choose for Δx, but usually either 0.01 or 0.001 is small enough.

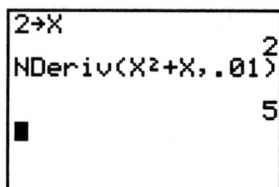

```
2→X
              2
NDeriv(X²+X,.01)
              5
■
```

The NUM Menu

Press the $\boxed{\text{MATH}}$ key again and use the cursor key to highlight the word NUM. NUM is an abbreviation for Number. It is followed by a four-item menu. The items are

1: Round(
2: IPart
3: FPart
4: Int

The first item, **Round(**, rounds a number, variable, expression, or matrix to the designated number of decimal places. The number, variable, expression, or matrix to be rounded and the designated number of places are separated by a comma. If the designated number of places is left out, the number is automatically rounded to ten digits. The usual rules for rounding are used. Suppose we wish to round 3.452 to one decimal place. The keystrokes are [MATH], [▷], [1], [3], [.], [4], [5], [2], [ALPHA], [.], [1], [)], and [ENTER]. The result is 3.5. If we wish to round 3.452 to two decimal places, the keystrokes are [MATH], [▷], [1], [3], [.], [4], [5], [2], [ALPHA], [.], [2], [)], and [ENTER]. The result is 3.45. If we wish to round 3.452 to a whole number, the number of decimal places is zero and the keystrokes are [MATH], [▷], [1], [3], [.], [4], [5], [2], [ALPHA], [.], [0], [)], and [ENTER]. The result is 3. If the quantity to be rounded is a matrix, then each entry in the matrix is rounded to the designated number of decimal places. For example, Round([A],0), rounds each entry in [A] to an integer.

```
Round(3.452,1)
               3.5
Round(3.452,2)
              3.45
Round(3.452,0)
                 3
```

The second item, **IPart**, finds the integer part of the number that follows the item. To find the integer part of -3.452, the keystrokes are [MATH], [▷], [2], [(−)], [3], [.], [4], [5], [2], and [ENTER]. The result is -3.

```
IPart -3.452
               -3
FPart -3.452
             -.452
■
```

The third item, **FPart**, finds the fractional part of the number that follows the item. To find the fractional part of -3.452, the keystrokes are [MATH], [▷], [3], [(−)], [3], [.], [4], [5], [2], and [ENTER]. The result is $-.452$.

```
Int -3.452
             -4
Int 3.452
              3
Int -3
             -3
■
```

The fourth item, **Int**, is the greatest integer function. It finds the greatest integer less than or equal to the number that follows the item. The keystrokes for finding the greatest integer in -3.452 are [MATH], [▷], [4], [(−)], [3], [.], [4], [5], [2], and [ENTER]. The result is -4, not -3, because -4 is less than -3.452. The keystrokes for finding the greatest integer in 3.452 are $\boxed{\text{MATH}}$, $\boxed{\triangleright}$, $\boxed{4}$, $\boxed{3}$, $\boxed{\cdot}$, $\boxed{4}$, $\boxed{5}$, $\boxed{2}$, and $\boxed{\text{ENTER}}$. The result is 3. The keystrokes for finding the greatest integer in -3 are $\boxed{\text{MATH}}$, $\boxed{\triangleright}$, $\boxed{4}$, $\boxed{(-)}$, $\boxed{3}$, and $\boxed{\text{ENTER}}$. The result is -3.

Problem Set 8

1. Evaluate the derivative of the function $f(x) = 2x^3 - 3x^2 - 36x + 25$ at $x = 0.5$.

2. Find the slope of the line tangent to $y = (3x^2 + 2)\sqrt{1 + 5x^2}$ at $x = 2$.

3. Use the keys from the MATH menu to find $\frac{5!}{2!3!}$ and $\frac{1759!}{1754!5!}$. Then, in paragraph form, discuss the results. State the largest $n!$ you can find using the calculator. Explain how to do a factorial problem if the calculator gives you an error statement.

4. For the number -81.2789, use the calculator to
   a. round to four significant digits.
   b. find the integer part of the number.
   c. find the fractional part of the number.
   d. find the greatest integer less than or equal to the number.
   e. Explain in paragraph form the uses for each of the four keys you used in parts a through d. Give specific examples illustrating the use of each key.

The $\boxed{\text{Y-VARS}}$ Key

Press the $\boxed{\text{2nd}}$ key followed by the $\boxed{\text{VARS}}$ key. On the first line are the names of three menus that can be accessed by highlighting the name of the menu. These are Y, ON, and OFF. Let's discuss each of them.

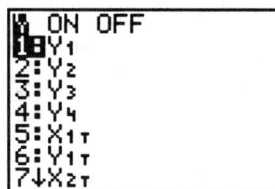

The Y Menu

Highlight the word Y. This ten-item menu has the items

1: Y_1
2: Y_2

3: Y₃ — rendered as Y_3
4: Y_4
5: X_{1T}
6: Y_{1T}
7: X_{2T}
8: Y_{2T}
9: X_{3T}
0: Y_{3T}

The first four items are the names of the functions available to us when the $\boxed{\text{MODE}}$ key is set for Functions. The last six are the names of the functions available to us when we are using the Parametric mode. Using this menu we can copy the name of a function into an expression. To do this, we press the $\boxed{\text{2nd}}$ key followed by the $\boxed{\text{VARS}}$ key, highlight the word Y, and press the number key associated with the item. The name of the item is copied into the expression. We can illustrate the use of this menu by graphing a circle such as $x^2 + y^2 = 4$. We have to solve the equation for y before we can graph it. The equation becomes $y = \pm\sqrt{4 - x^2}$. Press the $\boxed{\text{Y}=}$ key. The graph requires the use of both the Y_1 function and the Y_2 function. For Y_1 we press the $\boxed{\text{2nd}}$, $\boxed{x^2}$, $\boxed{(}$, $\boxed{4}$, $\boxed{-}$, $\boxed{\text{X|T}}$, $\boxed{x^2}$, $\boxed{)}$, and $\boxed{\text{ENTER}}$ keys, in that order. For Y_2 we can enter the same function preceded by a "minus" sign, or we can enter $-Y_1$ using $\boxed{(-)}$, $\boxed{\text{2nd}}$, $\boxed{\text{VARS}}$, $\boxed{1}$, and $\boxed{\text{ENTER}}$.

The ON and OFF Menus

Press the $\boxed{\text{2nd}}$ key followed by the $\boxed{\text{VARS}}$ key. If we highlight the words ON and then OFF, we see an eight-item menu for each. The items, which are identical except for the last word on each line, are

1: All − On	1: All − Off
2: Y_1 − On	2: Y_1 − Off
3: Y_2 − On	3: Y_2 − Off
4: Y_3 − On	4: Y_3 − Off
5: Y_4 − On	5: Y_4 − Off
6: X_{1T} − On	6: X_{1T} − Off
7: X_{2T} − On	7: X_{2T} − Off
8: X_{3T} − On	8: X_{3T} − Off

If we highlight the word OFF, press the number key associated with a particular function, and then press the $\boxed{\text{ENTER}}$ key, that function is turned off. In the previous menu, we discussed graphing the two

functions $Y_1 = \sqrt{4 - X^2}$ and $Y_2 = -Y_1$. If we turn off Y_2 by pressing the
$\boxed{3}$ key followed by the $\boxed{\text{ENTER}}$ key and try to graph the two functions,
only Y_1 will be graphed. If we press the $\boxed{Y=}$ key, the function is still
listed, but as far as the calculator is concerned the function does not exist,
because we have turned it off. If the Y_2 function is part of an expression
or a line in a program and we try to evaluate the expression or execute
the program, we will get an error message on the viewing screen. We can
turn the function back on by highlighting the word ON and again pressing
the $\boxed{3}$ key followed by the $\boxed{\text{ENTER}}$ key. On either menu, pressing the $\boxed{1}$
key followed by the $\boxed{\text{ENTER}}$ key turns all the functions on or off. When
we turn on or off one of the parametric functions, such as X_{1T}, the
corresponding Y function is also turned on or off.

The $\boxed{\text{DRAW}}$ Key

This key gives us access to menu items that let us draw lines, plot or
remove points, graph functions, and graph inequalities. Before using the
menu, we should make sure that the graphing modes we have set using
the $\boxed{\text{MODE}}$ key and the limits of the graph we have set using the
$\boxed{\text{RANGE}}$ key are appropriate. We should also check the $\boxed{Y=}$ key to
make sure that we have added the functions we want to graph and have
erased the functions we do not want to graph, because they will be drawn
along with the graphs created by this menu. Now press the $\boxed{\text{2nd}}$ key
followed by the $\boxed{\text{PRGM}}$ key. On the viewing screen is the word DRAW
highlighted and followed by a seven-item menu. The items in the menu
are

$$
\begin{aligned}
&\text{1: ClrDraw} \\
&\text{2: Line(} \\
&\text{3: PT} - \text{On(} \\
&\text{4: PT} - \text{Off(} \\
&\text{5: PT} - \text{Chg(} \\
&\text{6: DrawF} \\
&\text{7: Shade(}
\end{aligned}
$$

The first item, **ClrDraw**, clears the graphing screen of all drawings. To
execute the item, we press the $\boxed{1}$ key followed by the $\boxed{\text{ENTER}}$ key.
ClrDraw is copied onto a blank line on the viewing screen and is followed
by the word Done.

The second item, **Line(**, lets us draw a line between two points on a graph. The item can be executed in either of two ways. First, if we start from a graph on the viewing screen and execute the item by pressing the $\boxed{\text{2nd}}$, $\boxed{\text{PRGM}}$, and $\boxed{2}$ keys, the graph will again appear on the screen with blinking crosshairs at the origin. Using the cursor keys, we position the blinking crosshairs at the point where we want the line to begin and press the $\boxed{\text{ENTER}}$ key. The blinking crosshairs change to a blinking rectangle, and we use the cursor keys to position it at the point where we want the line to end. As we move the blinking rectangle, the line is displayed. Pressing the $\boxed{\text{ENTER}}$ key again draws the line on the graph between the two points. Second, if we begin from a blank line on the viewing screen and press the $\boxed{\text{2nd}}$, $\boxed{\text{PRGM}}$, and $\boxed{2}$ keys, **Line(** is copied onto that line. We then enter the x- and y-coordinates of the beginning point and the x- and y-coordinates of the ending point, separating the entries with commas. Press the $\boxed{)}$ and $\boxed{\text{ENTER}}$ keys. The line is drawn on the graph along with any functions on the Y = list.

For example, suppose we want to draw the line connecting the points $(-3, 2)$ and $(2, 4)$ on a graph. If we start from a graph, the keystrokes are $\boxed{\text{2nd}}$, $\boxed{\text{PRGM}}$, and $\boxed{2}$. We position the blinking crosshairs as close to the point $(-3, 2)$ as we can and press the $\boxed{\text{ENTER}}$ key. Then we move the blinking rectangle as close to the point $(2, 4)$ as possible and again press the $\boxed{\text{ENTER}}$ key. If we start from a blank line on the viewing screen, the keystrokes are $\boxed{\text{2nd}}$, $\boxed{\text{PRGM}}$, $\boxed{2}$, $\boxed{(-)}$, $\boxed{3}$, $\boxed{\text{ALPHA}}$, $\boxed{.}$, $\boxed{2}$, $\boxed{\text{ALPHA}}$, $\boxed{.}$, $\boxed{2}$, $\boxed{\text{ALPHA}}$, $\boxed{.}$, $\boxed{4}$, $\boxed{)}$, and $\boxed{\text{ENTER}}$. In either case, the line connecting the two points will be drawn.

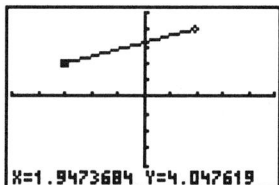

X=1.9473684 Y=4.047619

Line(-3,2,2,4)■

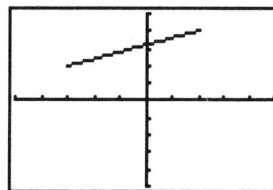

The third item, **PT − On(**, lets us plot a point on a graph. The item can be executed in either of two ways. First, if we start from a graph on the viewing screen and execute the item by pressing the $\boxed{\text{2nd}}$, $\boxed{\text{PRGM}}$, and $\boxed{3}$

keys, the graph will again appear on the screen with blinking crosshairs at the origin. Using the cursor keys, we position the blinking crosshairs at the location on the graph where we want the point to be and press the **ENTER** key. The point is drawn at that location. Second, if we begin from a blank line on the viewing screen and press the **2nd**, **PRGM**, and **3** keys, **PT − On(** is copied onto that line. We then enter the x- and y-coordinates of the point, separated by a comma. Press the **)** and **ENTER** keys. The point is drawn on the graph along with any functions on the $Y =$ list.

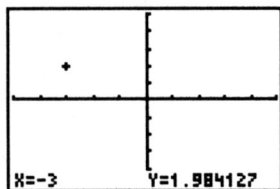

For example, suppose we want to plot the point $(-3, 2)$ on a graph. If we start from a graph, the keystrokes are **2nd**, **PRGM**, and **3**. We position the blinking crosshairs as close to the point $(-3, 2)$ as we can and press the **ENTER** key. If we start from a blank line on the viewing screen, the keystrokes are **2nd**, **PRGM**, **3**, **(−)**, **3**, **ALPHA**, **.**, **2**, **)**, and **ENTER**. In either case, the point is plotted on the graph. We can then press the **GRAPH** key to remove the crosshairs and see the point.

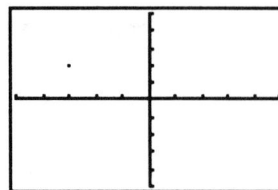

The fourth item, **PT − Off(**, lets us remove a point from a graph. The item can be executed in either of two ways. First, if we start from a graph on the viewing screen and execute the item by pressing the **2nd**, **PRGM**, and **4** keys, the graph will again appear on the screen with blinking crosshairs at the origin. Using the cursor keys, we position the blinking crosshairs at the location of the point on the graph and press the **ENTER** key. The point is erased. Second, if we begin from a blank line on the viewing screen and press the **2nd**, **PRGM**, and **4** keys, **PT − Off(** is copied onto that line. We then enter the x- and y-coordinates of the point, separated by a comma. Press the **)** and **ENTER** keys. The calculator graphs any functions on the $Y =$ list but erases the point from the graph.

For example, suppose we want to remove the point $(-3,2)$ from a graph. If we start from a graph, the keystrokes are 2nd, PRGM, and 4. We position the blinking crosshairs as close to the point $(-3,2)$ as we can and press the ENTER key. If we start from a blank line on the viewing screen, the keystrokes are 2nd, PRGM, 4, (−), 3, ALPHA, ⊡, 2,), and ENTER. In either case, the point is erased. Press the GRAPH key to see that the point has been removed.

The fifth item, **PT − Chg(**, lets us change the status of a point on a graph. If the point is plotted on a graph, this will erase it. If a point is not plotted on a graph, this will plot it. This item can be executed in either of two ways. First, if we start from a graph on the viewing screen and execute the item by pressing the 2nd, PRGM, and 5 keys, the graph will again appear on the screen with blinking crosshairs at the origin. Using the cursor keys, we position the blinking crosshairs at the location of the point on the graph and press the ENTER key. The status of the point is reversed. Second, if we begin from a blank line on the viewing screen and press the 2nd, PRGM, and 5 keys, **PT − Chg(** is copied onto that line. Enter the x- and y-coordinates of the point, separated by a comma. Press the) and ENTER keys. The calculator graphs any functions on the Y = list and reverses the status of this point.

For example, suppose we want to reverse the status of the point $(2,4)$ on a graph. If we start from a graph, the keystrokes are 2nd, PRGM, and 5. We position the blinking crosshairs as close to the point $(2,4)$ as we can and press the ENTER key. If we start from a blank line on the viewing screen, the keystrokes are 2nd, PRGM, 5, 2, ALPHA, ⊡, 4,), and ENTER. In either case, the status of the point is reversed as the calculator graphs the functions from the Y = list.

The sixth item, **DrawF**, lets us draw a function on a graph. The keystrokes are 2nd, PRGM, and 6. **DrawF** is copied onto the viewing screen. We then enter the name of the function, such as $x^2 - 4x$, and press the ENTER key. The function is graphed along with the functions from the Y = list. For example, the keystrokes for graphing $x^2 - 4x$ using this item are 2nd, PRGM, 6, X|T, x^2, ▦, 4, X|T, and

ENTER. Since the function we draw using this item is temporary, we cannot use the TRACE key on the function.

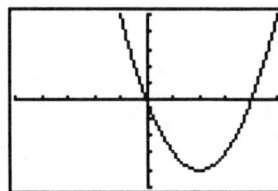

The seventh item, **Shade(**, lets us shade a specified portion of a graph. This shading can be below a function, above a function, or between two functions. If we press the 2nd, PRGM, and 7 keys, **Shade(** is copied onto the viewing screen. We must then follow this with a five-argument instruction, with each argument separated from the one that follows by a comma. After the last argument, we press the) and ENTER keys. Let's define each of the arguments.

The first argument defines the lower boundary of the shaded area. It can be a function such as $x^2 + 1$, a constant such as -4, or a function from the Y = list that is accessed by the Y menu on the Y-VARS key, such as **Y2**.

The second argument defines the upper boundary of the shaded area. It can be any of the types described for the first argument.

The third argument defines the resolution of the shading. It can be an integer from 1 to 8. It is optional, but if no value is specified, the resolution is the same as the setting for **Xres** on the RANGE key.

The fourth argument defines the left boundary of the shaded area (the beginning X). It can be a value, a variable, or an expression, but it cannot be a function. It is optional, but if no value is specified, the argument is the same as the setting for **Xmin** on the RANGE key.

The fifth argument defines the right boundary of the shaded area (the ending X). It can be a value, a variable, or an expression, but it cannot be a function. It is optional, but if no value is specified, the argument is the same as the setting for **Xmax** on the RANGE key.

This item is often used to graph inequalities. The inequalites must be solved for y in terms of x and must be in one of these forms:

$y \geq$ an expression in terms of x or a constant
$y >$ an expression in terms of x or a constant
$y \leq$ an expression in terms of x or a constant
$y <$ an expression in terms of x or a constant

These inequalities form the lower boundary of the shaded area, the first argument, if they are of the form $y \geq$ or $y >$. They form the upper boundary of the shaded area, the second argument, if they are of the form $y \leq$ or $y <$. Only the part of the inequality to the right of the inequality sign is included in the argument. If the inequality does not include the "=" sign, the boundary line should be a "dotted" line instead of a "solid" line, but the calculator cannot differentiate between the two.

If the third argument is 1, the area that fits the inequalities is solidly shaded. If the argument is 2 or greater, the area that fits the inequalities is shaded with vertical lines. The spacing between the vertical lines increases as the argument is increased. The third argument is optional, but it must be included if the fourth argument, the left boundary of the shaded area, is specified.

The fourth and fifth arguments are usually values. They are optional, but if the fifth argument is included, the third and fourth must also be included.

Now let's graph some examples. Use the RANGE key to set the minimum and maximum values for x and y at -10 and 10, respectively. Make sure that the Y $=$ list is clear of all functions.

Suppose we wish to shade the area bounded by the inequalities $y \leq x + 1$ and $y \geq x^2 - 4x$. The lower boundary will be $x^2 - 4x$ and the upper boundary will be $x + 1$. First, let's make sure the graphing screen is clear by pressing the 2nd, PRGM, 1, and ENTER keys. The keystrokes for entering the expression are [2nd], [PRGM], [7], [X|T], [x^2], [−], [4], [X|T], [ALPHA], [.], [X|T], [+], [1], [ALPHA], [.], [2], [)], and [ENTER].

```
Shade(X²-4X,X+1,
2)
```

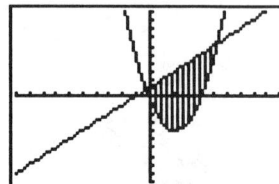

An alternate way to shade the area bounded by the inequalities $y \le x + 1$ and $y \ge x^2 - 4x$ is to store the inequalities as equations using the [Y=] key. For Y₁ the keystrokes are [X|T], [x^2], [−], [4], [X|T], and [ENTER]. For Y₂ the keystrokes are [X|T], [+], [1], and [ENTER]. Then we make sure the graphing screen is clear by pressing the [2nd], [PRGM], [1], and [ENTER] keys. The keystrokes for entering the expression are [2nd], [PRGM], [7], [2nd], [VARS], [1], [ALPHA], [.], [2nd], [VARS], [2], [ALPHA], [.], [2], [)], and [ENTER].

```
Shade(Y₁,Y₂,2)■
```

As a second example, let's shade the area bounded by the inequalities $3x + 2y \le 12$, $x \ge 0$, and $y \ge 0$. Clear the graphing screen by pressing the [2nd], [PRGM], [1], and [ENTER] keys. Solve the first inequality for y. It becomes $y \le 6 - \frac{3}{2}x$. This inequality is the upper boundary. The inequality $y \ge 0$ is the lower boundary, and the inequality $x \ge 0$ is the left boundary, so we must include an argument for the resolution. Let's use 2. The keystrokes are [2nd], [PRGM], [7], [0], [ALPHA], [.], [6], [−], [(], [3], [÷], [2], [)], [X|T], [ALPHA], [.], [2], [ALPHA], [.], [0], [)], and [ENTER]. We could have used $y \le 6 - 1.5x$ instead of $y \le 6 - \frac{3}{2}x$, but we wanted to illustrate that the upper and lower boundaries can include parentheses and/or fractions. This is useful if the fraction does not have a convenient decimal form.

```
Shade(0,6-(3/2)X
,2,0)■
```

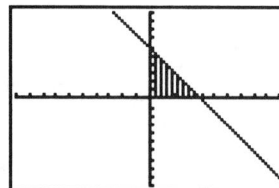

For the third example, let's shade the area bounded by the inequalities $x + 2y \leq 10$, $2x + y \leq 14$, $x \geq 0$, and $y \geq 0$. Clear the graphing screen by pressing the 2nd, PRGM, 1, and ENTER keys. Solving the first two inequalities for y, we have $y \leq 5 - 0.5x$ and $y \leq 14 - 2x$. The lower boundary is $y \geq 0$ and the left boundary is $x \geq 0$. Let's use 2 for the resolution. The upper boundary consists of the two inequalities $y \leq 5 - 0.5x$ and $y \leq 14 - 2x$. They can be put together to form the upper boundary using the technique for graphing functions with restricted domains discussed earlier, but we have to find the x value of their intersection point and also determine which function to include first. The x value of their intersection point can be found by setting $5 - 0.5x = 14 - 2x$ and solving for x. The solution is $x = 6$. We can determine which function to include first by comparing their y-intercepts and choosing the inequality with the smaller y-intercept. The y-intercept of $y \leq 5 - 0.5x$ is $y = 5$ and the y-intercept of $y \leq 14 - 2x$ is $y = 14$, so we start with $y \leq 5 - 0.5x$. The second argument will be $(5 - 0.5x)(x < 6) + (14 - 2x)(x \geq 6)$. The "=" part can be included on either part of the statement. The keystrokes are 2nd, PRGM, 7, 0, ALPHA, $\cdot$, ((, 5, $-$, 0, ▒, 5, X|T,)), ((, X|T, 2nd, MATH, 5, 6,)), $+$, ((, 1, 4, $-$, 2, X|T,)), ((, X|T, 2nd, MATH, 4, 6,)), ALPHA, $\cdot$, 2, ALPHA, $\cdot$, 0,)), and ENTER. If we wish to see the complete graph of the lines $y = 5 - 0.5x$ and $y = 14 - 2x$, we must enter them on the Y = list as **Y₁** and **Y₂**.

```
Shade(0,(5-0.5X)
(X<6)+(14-2X)(X≥
6),2,0)
```

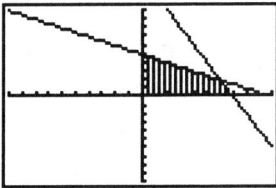

Problem Set 9

1. Write the procedure for using the **Shade(** item in the DRAW menu.

2. Use these instructions to explain, step-by-step, how to graph the system

$$x \geq 0$$
$$y \geq 0$$
$$2x + y \geq 4$$
$$x + 2y \geq 8$$

The VARS Key

Press the VARS key. On the first line are the names of five menus that

```
▓ Σ LR DIM RNG
1:n
2:x̄
3:Sx
4:σx
5:ȳ
6:Sy
7:σy
```

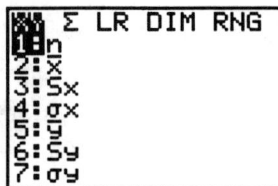

can be accessed by highlighting the name of the menu. These are XY, Σ, LR, DIM, and RNG. Let's discuss the RNG menu. We will discuss the first three in Chapter 3 and DIM in Chapter 4.

The RNG Menu

Use the cursor keys to highlight the word RNG. RNG is an abbreviation for the word Range. The items in this ten-item menu are

1: Xmin
2: Xmax
3: Xscl
4: Ymin
5: Ymax
6: Yscl
7: Xres
8: Tmin
9: Tmax
0: Tstep

The first seven items are the current settings for the Range variables for graphing in the Function mode. These were explained when we discussed the MODE key and the RANGE key earlier.

If we switch to the Parametric mode for graphing, the equations defining x and y are written in terms of the parametric variable t. At that time we have to define the smallest value for t that we will use, **Tmin**; the largest value for t that we will use, **Tmax**; and the increment between t values, **Tstep**. The last three items in the menu are the current settings for these variables.

We can display the value of an item by pressing the number key associated with the item and then the ENTER key. We can use the item in an expression, such as a line in a program, by pressing the number key associated with the item. The name of the item is copied into the expression. We can also do something with this menu that we cannot do with any of the other menus associated with the VARS key: we can use it to change the settings for the items. To do this, we enter the value we want for the item, press the STO key, and follow this with the VARS key, the ◁ key, and the number key associated with the item.

PROBLEMS USING GRAPHING APPLICATIONS

Problem Set 10

In Problems $1-5$, graph the three functions on the same set of axes. Use the trace and zoom to find the x- and y-intercepts. Write a paragraph explaining the similarities and relationships of the graphs. With examples, make predictions about similar graphs.

1. $f(x) = 0.3x$; $f(x) = 0.3x + 5$; $f(x) = 0.3x - 5$
2. $f(x) = \sin x$; $f(x) = 2 \sin x$; $f(x) = \sin 2x$
3. $f(x) = \tan x$; $f(x) = \tan\left(x + \frac{\pi}{2}\right)$; $f(x) = \tan\left(x - \frac{\pi}{2}\right)$
4. $f(x) = \ln x$; $f(x) = \ln(2x)$; $f(x) = \ln\left(\frac{1}{2}x\right)$
5. $f(x) = |x|$; $f(x) = \ln x + |x|$; $f(x) = \ln x - |x|$

In Problems $6-10$, find the limit of the function as x approaches the given value from the left and from the right. Discuss the continuity of the function at the given value.

6. $f(x) = \dfrac{x^2 - 3x + 4}{x - 5}$; $x \to 5$ 7. $f(x) = \dfrac{x^2 + x - 2}{x - 1}$; $x \to 1$

8. $f(x) = \dfrac{2x^2 - 5x + 2}{x^2 + x - 6}$; $x \to 2$ 9. $f(x) = \dfrac{x^4}{5^x}$; $x \to \infty$

10. $f(x) = \dfrac{3 - \sqrt{2x + 5}}{\sqrt{8x + 9} - 5}$; $x \to 2$

In Problems $11-17$, graph the function; find the relative maxima, relative minima, and inflection points correct to four decimal places; find the interval for each type of concavity; and find the asymptotes. Write a paragraph explaining how this is done using the graphing calculator.

11. $f(x) = \frac{1}{3}x^3 - x + 5$ 12. $f(x) = \dfrac{x^2 - 2x - 8}{x - 4}$

13. $f(x) = xe^{3x}$ 14. $f(x) = \ln(x^4 + x^2 + 1)$

15. $x^2 + 6xy + 25y^2 = 16$ 16. $y = \dfrac{x^2 + 1}{e^x}$

17. $e^{x^2} + e^{3y} = 9$

In Problems $18 - 21$, use a Taylor series to approximate the given function $y = f(x)$ at $x = c$ for the given degree n.

18. $f(x) = \cos 2x$; $x = 0$; $n = 3$ 19. $f(x) = \tan x$; $x = \frac{\pi}{4}$; $n = 4$

20. $f(x) = \ln x$; $x = 0$; $n = 6$ 21. $f(x) = e^{\ln x}$; $x = 1$; $n = 6$

In Problems $22 - 25$, find the numerical value, correct to four decimal places, of the derivative of the function at the given point. Write a paragraph explaining what you have found.

22. $y = \dfrac{e^{3x}}{e^{3x} + 1}$ at $x = 1.75$ 23. $y = \dfrac{x^2 + 5x - 6}{x^2 - 1}$ at $x = 1$

24. $y = \dfrac{2^x - 1}{\ln x}$ at $x = 3.45$

25. $y = \dfrac{e^{x^2 - x + 1} \ln (x^3 - 3x^2 + 4x + 3)}{(x - 2)\sqrt{1 - x^2}}$ at $x = 0.5$

In Problems $26 - 27$, let n be an integer such that $-10 \le n \le 10$.

26. Make a table for n^3.

27. Make a table, rounded to five decimal places, for $\sqrt[3]{n}$.

Using the graphing calculator, find the integer and fractional part of each entry in the table.

28. Make tables showing the values for $\dfrac{n!}{(n - r)!}$ and $\dfrac{n!}{(n - r)!r!}$ where $n = 20$ and r is a whole number less than or equal to 20.

In Problems $29 - 33$, graph the piecewise defined functions.

29. $f(x) = \begin{cases} x^2 - 7 & \text{for } 0 \le x \le \sqrt{7} \\ x^{1/2} & \text{for } \sqrt{7} < x \le 16 \\ 4 & \text{for } x > 16 \end{cases}$

30. $f(x) = \begin{cases} 2 + \sqrt{x^2 + 1} & \text{for } 0 \le x \le 7 \\ \sqrt{x^2 + 2} & \text{for } 7 < x \le 9 \\ 10 & \text{for } x > 9 \end{cases}$

31. $f(x) = \begin{cases} (x + 1)^2 & \text{for } -3 \le x < -1 \\ x + 1 & \text{for } -1 \le x < 0 \\ -x + 1 & \text{for } 0 \le x < 1 \\ (x - 1)^2 & \text{for } 1 \le x \le 3 \end{cases}$

32. $f(x) = \begin{cases} \ln|x+1| & \text{for } 0 \le x \le 5 \\ e^{-0.0001x} & \text{for } 5 < x \le 8 \\ \sqrt{x} & \text{for } x > 8 \end{cases}$

33. $f(x) = \begin{cases} \dfrac{x+2}{x-3} & \text{for } 0 < x < 3 \text{ or } x > 3 \\ 8 & \text{for } x = 3 \end{cases}$

In Problems $34 - 39$, find the solution set by graphing the system of inequalities.

34. $\begin{cases} x + 2y \le 40 \\ 3x + y \le 45 \end{cases}$

35. $\begin{cases} 3x + 2y \le 17 \\ x + 4y \ge 9 \end{cases}$

36. $\begin{cases} y > 1 - x \\ y > 2x + 1 \end{cases}$

37. $\begin{cases} 3x + 2y \le 4 \\ 2x - 3y \ge 9 \end{cases}$

38. $\begin{cases} x \ge 0 \\ y \ge 0 \\ y \ge x + 1 \\ y \le \dfrac{17 - 3x}{2} \end{cases}$

39. $\begin{cases} x \ge 0 \\ y \ge 0 \\ y \le x + 1 \\ y \ge \dfrac{17 - 3x}{2} \end{cases}$

In Problems $40 - 44$, use a Maclaurin series to approximate the function to nine decimal places.

40. $\sin 29°$

41. $\sqrt[3]{e}$

42. $\ln 1.04$

43. $\tan 46°$

44. $\sqrt{5.02}$

In Problems $45 - 50$, find the maximum or minimum as indicated.

45. An open, rectangular box is to be constructed from a 6-foot by 8-foot piece of 28-gauge sheet metal by cutting out equal-sized squares from each corner, folding up the sides, and soldering the seams. Find the dimensions of the box with the maximum volume.

46. The BXG Corporation is going to tunnel to a position where there is thought to be a new gold deposit. From the designated starting point, the deposit is thought to be 2000 feet down and 6000 feet to the west. A vertical core was drilled, and it was found that there is 1000 feet of packed earth on top of bedrock. If it costs $50 per linear foot to tunnel through packed earth and $110 per linear foot to tunnel through bedrock, what is the minimum cost to tunnel to the gold deposit?

47. If money moves through the European and United States banking systems to the open market according the the formula, $s = \sqrt{5t} + \dfrac{10}{\sqrt{5t}}$, what is the optimum time (t) in days for a supply of money (s) to reach the open market?

48. You have been hired by CILIZ's Construction Company as the company's CPA. In reorganizing your new office, you are going to have to move computer systems through a passageway into a corridor that is at a right angle to the passageway. The systems are 3 feet wide and vary in length. The passageway is 8 feet wide and the corridor is 14 feet wide. What is the length, in feet, of the longest computer system that can make the turn?

49. You are designing a cylindrical can that is to hold a quart of liquid. The material for the top and base of the can costs 50¢ per square inch and material for the side costs 40¢ per square inch. If the cost is to be minimized, what are the dimensions of the can?

50. A service organization charges $50 per subscriber, less $0.50 for each subscriber over 500 or plus $0.50 for each subscriber under 500. How many subscribers would maximize the revenue? What is that revenue?

Chapter 2   PROGRAMMING CAPABILITIES

Programming can be a useful tool for repeatedly performing operations that involve several steps. Graphing derivatives of functions and finding the roots using Newton's method are examples of such operations. Programming allows us to store the steps involved as a program in the calculator. We can then access the program whenever we need to perform the operation. Before actually writing and storing a program in the calculator, let's examine the various keys we will need to use.

USING THE GRAPHING CALCULATOR

The PRGM Key

```
EXEC EDIT ERASE
1:Prgm1
2:Prgm2
3:Prgm3
4:Prgm4
5:Prgm5
6:Prgm6
7↓Prgm7
```

Press the PRGM key. PRGM is an abbreviation for the word Program. On the first line of the viewing screen are the words EXEC, EDIT, and ERASE, with the word EXEC highlighted. EXEC is an abbreviation for the word Execute. The calculator has a total of 2400 bytes of memory available and can store up to 37 programs. The programs are accessed by the numbers 1 through 0, the letters A through Z, and the Greek letter θ. The lines that follow the first line on the viewing screen list the 37 available programs. We can access a program by entering its number or letter. If the program name is a letter, we access that program by pressing the ALPHA key followed by the key that has the desired letter to the right and above it. If we wish to access PrgmQ, we press the ALPHA key followed by the 9 key. θ is to the right and above the 3 key. We can also use the ∇ and △ keys to scroll through the list of programs and press the ENTER key when the correct number or letter is highlighted. When we store a program in the calculator, we start by giving the program a title. For example, if we wish to write a program that graphs the derivatives of functions, we might use the title GRAPHDER. The calculator lists that title to the right of the program number or letter. We access the program by the number or letter, but the title helps us remember what the program does. If Program 1 graphs the derivatives of functions, then the first program line would read **1:Prgm1 GRAPHDER.** If there is no title after the program number or letter, that program is not in use.

The EXEC Function

When EXEC is highlighted, we can execute or run a program. We tell the calculator which program we wish to run by entering the number or letter of the program, or by highlighting the number or letter and pressing the $\boxed{\textbf{ENTER}}$ key. The program name is copied onto the Home screen and we begin execution by pressing the $\boxed{\textbf{ENTER}}$ key.

The ERASE Function

When ERASE is highlighted, the program title is replaced with the number of bytes of memory the program uses. When we select the number or letter of the program we wish to erase, the ERASE program menu appears. The first line lists the program name and title. Below this are two menu selections. We can press the $\boxed{\textbf{1}}$ key if we decide not to erase the program, or the $\boxed{\textbf{2}}$ key if we wish to erase the program. Note that 1: is highlighted automatically so that we do not accidentally erase a program we wish to keep. After either selection we are returned to the Home screen.

The EDIT Function

When EDIT is highlighted, we can store a new program in the calculator or make changes in an existing program. Highlight the word EDIT and choose the name of a program that is not in use. Let's assume the program is **Prgm1.** Press the $\boxed{\textbf{1}}$ and $\boxed{\textbf{ENTER}}$ keys. On the first line of the screen we have the program name, Prgm1:, followed by a special blinking rectangle. The calculator wants us to enter the title of the program. The special blinking rectangle means that the calculator expects the title to consist of letters, so it has set itself to receive letters automatically. This means that we do not have to press the $\boxed{\textbf{ALPHA}}$ key first to enter the letter. If we wish to enter the letter G, we just press the

```
Prgm1:
:
```

TAN key. The title can have up to 8 letters. If the title has fewer than 8 letters, we must tell the calculator we are finished by pressing the **ENTER** key. For now, let's just press the **ENTER** key. The blinking rectangle changes to its usual form and moves to the first line of the program, which begins with a colon. A program is a series of steps that are normally executed sequentially. Each step includes one or more instructions. These instructions can take the form of an algebraic expression to evaluate, but often they are in the form of a logical decision to be made by the calculator or a form to receive input from us or to display output on the viewing screen. Each step in a program begins with a colon. The colon is placed on the viewing screen automatically by the calculator. The instructions in a step can use more than one line on the screen. We tell the calculator that the instructions for a step are finished by pressing the **ENTER** key. Before we are ready to enter program instructions, we need to examine three more menus. Press the **PRGM** key again. The first of the three program instruction menus should appear on the screen. The first line has the words CTL, I/O, and EXEC, with CTL highlighted. When we select an instruction from the CTL or I/O menus, that instruction is copied to the current line of the program. When we select a program name from the EXEC menu, that name is copied to the current line of the program. Since understanding how to use these menus is a major part of programming, let's examine each of them separately. CTL is highlighted, so let's examine this menu first.

```
CTL I/O EXEC
1:Lbl
2:Goto
3:If
4:IS>(
5:DS<(
6:Pause
7↓End
```

The CTL Menu

CTL is an abbreviation for the word Control. The following eight instructions control how a program is executed. They change the normal step-by-step sequential execution of a program. The items are

$$1: \text{Lbl}$$
$$2: \text{Goto}$$
$$3: \text{If}$$
$$4: \text{IS>(}$$
$$5: \text{DS<(}$$
$$6: \text{Pause}$$
$$7: \text{End}$$
$$8: \text{Stop}$$

The first two instructions, **Lbl** and **Goto**, are used together to transfer program execution to a particular step in the program. Each of them must be followed by a number or letter. For example, if the calculator encounters the instruction Goto 3, it transfers control to the step in the program that has the instruction Lbl 3 and continues execution from that line.

The **If** instruction is used for testing the relationship between two variables. Transfer of control is determined by the result of the test. **If** is followed by an expression such as $A > B$, $A \geq B$, $A < B$, $A \leq B$, $A = B$, or $A \neq B$. Either of the variables may be replaced by a number. Suppose the instruction is: If $A > 0$. If the instruction is true, execution continues with the next step in the program. If the instruction is not true, the next step is skipped, and execution continues with the second step after the instruction.

The **IS>(** instruction (IS stands for increment and skip) must be followed by two arguments: a variable and an expression or number. The two arguments must be separated by a comma and the second is followed by a close parenthesis. When the calculator executes the step containing the instruction, it evaluates the second argument and adds 1 to the value of the first variable. If the value of the first variable is less than or equal to the value of the second expression or number, execution continues with the next step in the program. If the value of the first variable is greater than the value of the second expression or number, the next step is skipped and execution continues with the second step after the instruction.

The **DS < (** instruction (DS stands for decrement and skip) is exactly like the **IS>(** instruction except that 1 is subtracted from the value of the first variable and the next step is skipped if the value of the first variable is less than the value of the second expression or number.

The **Pause** instruction stops execution of the program until we press the ENTER key. This allows us time to examine results or view graphs that are displayed on the viewing screen. When we press the ENTER key, execution continues with the next step in the program.

The **End** instruction lets the calculator know that we are finished executing the program. If we forget to put this instruction at the end of a program, the calculator assumes we are finished, since the program has no more steps to execute.

The **Stop** instruction is the last one on the list of the CTL instructions. To see it on the viewing screen, we must use the $\boxed{\nabla}$ key and scroll down the list of instructions. The **Stop** and **End** instructions are similar. The **Stop** instruction, however, returns us to the Home screen; the **End** instruction does not.

The I/O Menu

Use the $\boxed{\triangleright}$ key and highlight I/O. I/O is an abbreviation for Input/Output. The following five instructions control our input to the program and the output from a program to the viewing screen. The items are

```
CTL I/O EXEC
1:Disp
2:Input
3:DispHome
4:DispGraph
5:ClrHome
```

> 1: Disp
> 2: Input
> 3: DispHome
> 4: DispGraph
> 5: ClrHome

The **Disp** instruction allows us to display a text message or the current value of a variable on the viewing screen. The text message must be preceded by a quote ("). A closing quote is optional. The instruction with a text message might be Disp "X=" or Disp "AREA IS". To enter the text we must use the $\boxed{\text{ALPHA}}$ key with each letter of the text, or we can press the $\boxed{\text{2nd}}$ key followed by the $\boxed{\text{ALPHA}}$ key. This locks the keyboard of the calculator into the alphabetic mode. When a key is pressed, the calculator automatically puts the letter associated with the key on the screen. For example, if we press the $\boxed{\text{TAN}}$ key, the calculator puts a G on the screen. We can enter a space by pressing the $\boxed{0}$ key and a question mark by pressing the $\boxed{(-)}$ key. To remove the lock, we have to press the $\boxed{\text{ALPHA}}$ key again. To display the current value of a variable, we follow the **Disp** instruction with the name of the variable. An example of this is Disp X.

The **Input** instruction, when followed by a variable, allows us to input data into the program during execution. The calculator will prompt us with a question mark (?) when data are required. After entering the data, we must press the ENTER key to resume execution. An example of this type of instruction is Input X. If we wish to be reminded which variable we are entering, we must precede this instruction with a step that has an instruction such as Disp "X=" or Disp "ENTER A VALUE FOR X". If we use the **Input** instruction without a following variable, the calculator automatically displays the current graph. We can use the cursor to find the values of the X and Y variables on the graph. To resume execution, we must press the ENTER key.

The **DispHome** instruction displays the Home screen from a program.

The **DispGraph** instruction displays the current graph. However, there is no cursor on the graph, so we cannot use the TRACE key.

The **ClrHome** clears the Home screen from a program.

Problem Set 1

Use the following program to answer Problems 1 − 3.

```
:0→I
:0→T
:Lbl 1
:Disp "X="
:Input X
:T+X→T
:I+1→I
:If I < 5
:Goto 1
:T/I→T
:Disp T
:End
```

1. Describe in paragraph form what the program does. Include a description of the function of each step in the program.
2. Rewrite the program using the **IS > (** instruction. Describe how it changes the program.
3. Rewrite the program using the **DS < (** instruction. Describe how it changes the program.

The EXEC Menu

```
CTL I/O EXEC
1▉Prgm1
2:Prgm2
3:Prgm3
4:Prgm4
5:Prgm5
6:Prgm6
7↓Prgm7
```

Use the ▷ key and highlight EXEC. On the viewing screen are the names of the first seven of the 37 programs and the titles of the ones in use. We can select the number or letter of a program. The calculator copies the name of the program to the current line of the program we are writing or editing. When the calculator executes this line of the program, it treats this program as a subroutine. That is, it transfers control to this new program, executes it, and then transfers control back to the next line of the original program and continues execution of the original program.

EXAMPLES OF PROGRAMMING CAPABILITIES

A Program for Graphing the Derivative of a Function

If we store a function in the calculator as Y1 using the $\boxed{Y=}$ key, this program graphs the function on the viewing screen, pauses to allow us to examine it, and then graphs the derivative of the function. It does not give us the equation of the derivative, only its graph. This program uses items from menus associated with the $\boxed{\text{VARS}}$, $\boxed{\text{MATH}}$, $\boxed{\text{Y-vars}}$, and $\boxed{\text{DRAW}}$ keys discussed in Chapter 1.

Now let's write the program and store it in the calculator. We will write the lines as they will appear in the program. Press the $\boxed{\text{PRGM}}$ key and use the ▷ key to highlight EDIT. Let's assume the program is to be stored in Prgm 1, so press the $\boxed{1}$ key or the $\boxed{\text{ENTER}}$ key since the 1 is highlighted. The calculator responds with Prgm1: and waits for us to give the program a title. Let's call it GRAPHDER. The calculator expects the title to be alphabetic rather than numerical, so we do not have to press the $\boxed{\text{ALPHA}}$ key before entering the letters in the title. The keystrokes are $\boxed{\text{TAN}}$, $\boxed{\times}$, $\boxed{\text{MATH}}$, $\boxed{8}$, $\boxed{\wedge}$, $\boxed{x^{-1}}$, $\boxed{\text{SIN}}$, and $\boxed{\times}$. We do not have to press the $\boxed{\text{ENTER}}$ key because the title has eight letters. Now the blinking rectangle on the calculator is at the beginning of the first line of the program after the colon (:) and we are ready to start entering instructions. After each line of the program, we will list the keystrokes necessary to create the line and make some comments about the line.

```
Prgm1:GRAPHDER
:▉
```

: ClrDraw

The keystrokes are 2nd, PRGM, 1, and ENTER. The 2nd, PRGM and 1 keystrokes copy the word ClrDraw onto the line on the viewing screen. The item was explained when we discussed the DRAW key in Chapter 1. This instruction erases any graphs from the viewing screen before we graph the new function. The ENTER key tells the calculator we are finished with this line of instruction.

: 95→N

The keystrokes are 9, 5, STO, LOG, and ENTER. When we press the STO key, the calculator assumes the next keystroke will be the name of a variable, so we do not have to press the ALPHA key before entering the letter. This line determines the number of points the calculator will use in drawing the graph of the derivative.

: DispGraph

```
Prgm1:GRAPHDER
:ClrDraw
:95→N
:DispGraph
:Pause
:(Xmax-Xmin)/N→S

:1→I
```

The keystrokes are PRGM, ▷, and 4 (to copy the word DispGraph onto the line) and ENTER. This line draws the graph of the function, which we stored as Y_1, on the viewing screen.

: Pause

The keystrokes are PRGM and 6 (to copy the word Pause onto the line) and ENTER. This instruction stops execution of the program and allows us to view the graph. It is an optional line. If we did not include it, the calculator would immediately start graphing the derivative.

: (Xmax − Xmin)/N→S

The keystrokes are (, VARS, ◁, 2, −, VARS, ◁, 1,), ÷, ALPHA, LOG, STO, LN, and ENTER. The VARS, ◁, 2 and VARS, ◁, 1 keystrokes copy the words Xmax and Xmin onto the line. These items in the RNG menu of the VARS key allow us to access the Range variables stored for Xmax and Xmin. This line determines the increment between the points used in the graph of the derivative.

: 1→I

The keystrokes are 1, STO, x^2, and ENTER. We will use the IS>(

instruction as a counter to keep track of the number of points we wish to calculate. This instruction starts the increment part of that counter at 1.

: Xmin→A

The keystrokes are $\boxed{\text{VARS}}$, $\boxed{◁}$, $\boxed{1}$, $\boxed{\text{STO}}$, $\boxed{\text{MATH}}$, and $\boxed{\text{ENTER}}$. This instruction sets the *x*-coordinate of the first point, which we call A, as the Range variable Xmin.

: A→X

The keystrokes are $\boxed{\text{ALPHA}}$, $\boxed{\text{MATH}}$, $\boxed{\text{STO}}$, $\boxed{\text{X|T}}$, and $\boxed{\text{ENTER}}$. This instruction stores the value of A in the variable X. In the next instruction we are going to calculate the *y*-coordinate of the first point. That calculation is done by substituting the value of X into the function stored as Y_1, so we need to make sure X has the right value.

: NDeriv(Y_1,.001)→B

```
:Xmin→A
:A→X
:NDeriv(Y₁,.001)
→B
:Lbl 1
:X+S→X
:NDeriv(Y₁,.001)
→Y■
```

The keystrokes are $\boxed{\text{MATH}}$, $\boxed{8}$, $\boxed{\text{2nd}}$, $\boxed{\text{VARS}}$, $\boxed{1}$, $\boxed{\text{ALPHA}}$, $\boxed{.}$, $\boxed{.}$ $\boxed{0}$, $\boxed{0}$, $\boxed{1}$, $\boxed{)}$, $\boxed{\text{STO}}$, $\boxed{\text{MATRIX}}$, and $\boxed{\text{ENTER}}$. The $\boxed{\text{MATH}}$ and $\boxed{8}$ keystrokes copy the word NDeriv(onto the line. It was explained in Chapter 1 when we discussed the $\boxed{\text{MATH}}$ key. This instruction calculates an approximation for the numerical value of the derivative of the function and stores the value in variable B. It is the *y*-coordinate of the point (A,B).

: Lbl 1

The keystrokes are $\boxed{\text{PRGM}}$, $\boxed{1}$, $\boxed{1}$, and $\boxed{\text{ENTER}}$. The first two keystrokes copy the word Lbl onto the line. This instruction is used in conjunction with the Goto instruction which we will include later. When the calculator encounters the Goto instruction, it transfers execution of the program to the corresponding Lbl instruction.

: X + S→X

The keystrokes are $\boxed{\text{X|T}}$, $\boxed{+}$, $\boxed{\text{ALPHA}}$, $\boxed{\text{LN}}$, $\boxed{\text{STO}}$, $\boxed{\text{X|T}}$, and $\boxed{\text{ENTER}}$. This instruction increments the value of X by the amount S which was calculated earlier and stores it in the variable X.

: NDeriv(Y₁,.001)→Y

The keystrokes are **MATH**, **8**, **2nd**, **VARS**, **1**, **ALPHA**, **.**, **.** **0**, **0**, **1**, **)**, **STO**, **1**, and **ENTER**. This instruction calculates another approximation for the numerical value of the derivative using the new value for X and stores it in variable Y. This instruction and the one preceeding create a second point (X,Y).

: Line(A,B,X,Y)

The keystrokes are **2nd**, **PRGM**, **2**, **2nd**, **ALPHA**, **MATH**, **.**, **MATRIX**, **.**, **X|T**, **.**, **1**, **ALPHA**, **)**, and **ENTER**. The **2nd**, **PRGM**, and **2** keystrokes copy the word Line(onto the line. This item was explained when we discussed the **DRAW** key in Chapter 1. The **2nd** and **ALPHA** keystrokes put the calculator into the alphabetic mode so that we can enter the letters and commas that follow without having to press the **ALPHA** key before each entry. The last **ALPHA** keystroke changes the calculator back to the numerical mode. This instruction draws a line connecting points (A,B) and (X,Y).

: X→A
: Y→B

```
:NDeriv(Y₁,.001)
→Y
:Line(A,B,X,Y)
:X→A
:Y→B
:IS>(I,N)
:Goto 1
:End■
```

The keystrokes for the first line are **X|T**, **STO**, **MATH**, and **ENTER**. The keystrokes for the second line are **ALPHA**, **1**, **STO**, **MATRIX**, and **ENTER**. These two lines transfer the coordinates of point (X,Y) to point (A,B). We will calculate new values for the point (X,Y) and draw another line connecting these two points. Since we set the variable N to be 95 at the beginning of the program, the calculator will draw 95 of these lines to create the graph of the derivative.

: IS > (I, N)

The keystrokes for this line are **PRGM**, **4**, **2nd**, **ALPHA**, **x²**, **.**, **LOG**, **ALPHA**, **)**, and **ENTER**. The first two keystrokes copy the word IS > (onto the line. This is the instruction for incrementing the counter I by 1. The calculator does this and then compares it with the value of the variable N. If I is less than or equal to N, the calculator executes the next line of the program. If I is greater than N, the calculator skips over the next line and executes the second line following this instruction.

: Goto 1

The keystrokes for this line are $\boxed{\text{PRGM}}$, $\boxed{2}$, $\boxed{1}$, and $\boxed{\text{ENTER}}$. The first two keystrokes copy the word Goto onto the line. This instruction transfers execution of the program to the line which reads "Lbl 1" so that the calculator can calculate the next point (X,Y).

: End

The keystrokes for this line are $\boxed{\text{PRGM}}$ and $\boxed{7}$. We do not have to press the $\boxed{\text{ENTER}}$ key. This line is optional. The calculator automatically places an End statement as the last line of the program if we do not include it.

Before executing this program, we must store the function as Y₁ using the $\boxed{\text{Y}=}$ key. Reset the Range variables, if necessary. During execution when the calculator is graphing the function on the viewing screen, there is a small dark rectangle in the upper right corner of the screen. When the graph is finished, the rectangle disappears. The calculator is then waiting for us to press the $\boxed{\text{ENTER}}$ key to continue execution.

A Program for Finding the Roots of a Function

This program uses Newton's method for estimating the roots of a function. We must store the function as Y₁, using the $\boxed{\text{Y}=}$ key, and give the calculator our estimate of the root. The calculator refines that estimate correct to nine decimal places. In the following program, we will list the keystrokes for and comment on only the instructions that are different from the ones discussed previously.

: NEWTON
: Lbl 4
: Disp "ROOT ESTIMATE = "
: Input X

The keystrokes for the third line are $\boxed{\text{PRGM}}$, $\boxed{\triangleright}$, $\boxed{1}$, $\boxed{\text{2nd}}$, $\boxed{\text{ALPHA}}$, $\boxed{+}$, $\boxed{\times}$, $\boxed{7}$, $\boxed{7}$, $\boxed{4}$, $\boxed{0}$, $\boxed{\text{SIN}}$, $\boxed{\text{LN}}$, $\boxed{4}$, $\boxed{x^2}$, $\boxed{\div}$, $\boxed{\text{MATH}}$, $\boxed{4}$, $\boxed{\text{SIN}}$, $\boxed{\text{2nd}}$, $\boxed{\text{MATH}}$, $\boxed{1}$, $\boxed{\text{2nd}}$, $\boxed{\text{ALPHA}}$, $\boxed{+}$, and $\boxed{\text{ENTER}}$. The keystrokes for the fourth line are $\boxed{\text{PRGM}}$, $\boxed{\triangleright}$, $\boxed{2}$, $\boxed{\text{X|T}}$, and $\boxed{\text{ENTER}}$. The third line tells us what we are being asked to enter when the fourth line is executed. The fourth line puts a question mark on the viewing screen and waits for us to enter a value for X.

> : Lbl 1
> : If NDeriv(Y_1,.001) = 0
> : Goto 3

The second and third lines check to see if the slope of the tangent line is zero. If it is, execution is transferred to Lbl 3.

> : $Y_1 \rightarrow Y$
> : $X - Y/NDeriv(Y_1,.001) \rightarrow X$

This instruction calculates a new estimate for the root of the function and stores it in the variable X.

> : If abs $Y_1 < 1\text{E}-10$

The keystrokes for this line are $\boxed{\text{PRGM}}$, $\boxed{3}$, $\boxed{\text{2nd}}$, $\boxed{x^{-1}}$, $\boxed{\text{2nd}}$, $\boxed{\text{VARS}}$, $\boxed{1}$, $\boxed{\text{2nd}}$, $\boxed{\text{TEST}}$, $\boxed{5}$, $\boxed{1}$, $\boxed{\text{EE}}$, $\boxed{(-)}$, $\boxed{1}$, $\boxed{0}$, and $\boxed{\text{ENTER}}$.

> : Goto 2
> : Goto 1
> : Lbl 2
> : Disp "ROOT IS"
> : Disp X
> : END
> : Lbl 3
> : Disp "SLOPE IS ZERO"
> : Disp "CHOOSE ANOTHER"
> : Disp "ESTIMATE"
> : Goto 4

Newton's method is an iterative method for calculating the slope of the tangent line to the curve at a chosen point and uses the equation of this tangent line to refine the estimate. In successive estimates, the chosen point will be the x-coordinate of the x-intercept of the tangent line found in the previous estimate. This process is repeated until the desired accuracy is achieved. If the slope of the tangent line is zero, the method will not work. For example, the function $y = x^2 + 8x + 15$ has roots at $x = -5$ and -3. If we choose as our estimate $x = -4$, the slope of the tangent line is zero. The program checks for this and directs us to choose another estimate.

The procedure does not always locate the root closest to the estimate. The root located depends on the slope of the tangent line. For example, the function $y = x^3 + 2x^2 - x - 2$ has roots of -2, -1, and 1. In the graph at the left, we chose $x = 0.1$ as the first estimate. The slope of the tangent line at that point is -0.57, and the equation of the tangent line is $y = -0.57x - 2.022$. This line has an x-intercept of -3.5474, to four decimal places. The root located is the one at $x = -2$.

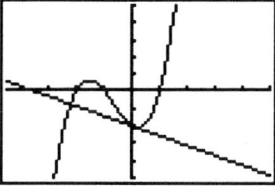

A Program for Drawing Tangent Lines to a Function

This program graphs a function that we have stored as Y1, using the $\boxed{Y=}$ key. At selected points on the function, it draws tangent lines to the function and plots the value of the derivative as a point. Execution of the program is terminated when we select a point that is larger than the Range variable which we have stored as Xmax. In the following program, we will list the keystrokes for and comment on only the instructions that are different from the ones discussed previously.

```
: TANGENT
: ClrDraw
: DispGraph
: Pause
: Disp "POINT IS"
: Input X
: Lbl 1
: NDeriv(Y1,.001)→M
: PT − On(X,M)
```

The eighth line of the program stores the numerical value of the derivative in the variable M. The value is the slope of the tangent line to the function at the point (X,Y). The keystrokes for the ninth line of the program are $\boxed{2\text{nd}}$, $\boxed{\text{PRGM}}$, $\boxed{3}$, $\boxed{\text{X|T}}$, $\boxed{\text{ALPHA}}$, $\boxed{.}$, $\boxed{\text{ALPHA}}$, $\boxed{\div}$, $\boxed{)}$, and $\boxed{\text{ENTER}}$. The instruction turns on the point whose coordinates are (X,M). The y-coordinate of the point gives us an approximation of the value of the derivative at the point X.

$$: Y_1 \rightarrow Y$$
$$: X - 5 \rightarrow A$$
$$: Y - 5M \rightarrow B$$
$$: X + 5 \rightarrow C$$
$$: Y + 5M \rightarrow D$$
$$: \text{Line}(A,B,C,D)$$

Lines two through five calculate the coordinates of the line drawn connecting points (A,B) and (C,D). The 5, an arbitrary choice, makes the line long enough to extend beyond the limits of the viewing screen.

$$: \text{Pause}$$
$$: \text{Disp "POINT IS"}$$
$$: \text{Input X}$$
$$: \text{If } X < \text{Xmax}$$
$$: \text{Goto 1}$$
$$: \text{End}$$

Problem Set 2

1. Graph the derivative of the function $y = \dfrac{3x^3}{\sqrt[5]{x^2}} - \dfrac{7x}{\sqrt[3]{x^4}} + 8\sqrt[7]{x^3}$.

2. Find the roots of the function $f(x) = \dfrac{2x+3}{10-x} - \dfrac{2x}{25-3x} + \dfrac{13}{2}$ using Newton's method.

3. Draw line tangent to the graph of $x^2 + y^2 = 58$ at $(7,3)$.

A Program for Numerical Integration Using Simpson's Rule

Before executing this program, we must store the functions as Y₁ and Y₂ using the $\boxed{Y=}$ key. The upper function is stored as Y₁. If either of the functions is the x-axis, it is stored as $Y = 0$. We must also store the beginning and ending points for calculating the area in variables A and B, respectively. To store the points, we enter the value of the point, press the STO key, the key for the variable, and the ENTER key. The program starts execution by asking us the number of intervals we want to use in approximating the area. The number must be an even number. The sixth line of the program checks to make sure we have entered an even number and has us re-enter the number if it is not even. The larger the number, the greater the accuracy of the approximation, but execution takes longer. The program graphs the function, shades the area we are approximating, and then displays the value of the area. It pauses after graphing the function to give us time to view the area. In the following program, we will list the keystrokes for and comment on only the instructions that are different from the ones discussed previously.

: SIMPRULE
: ClrDraw
: Lbl 4
: Disp "INTERVALS = "
: Input N
: If IPart(N/2) = N/2

The keystrokes for this line are PRGM, 3, MATH, ▷, 2, ((, ALPHA, LOG, ÷, 2,)), 2nd, MATH, 1, ALPHA, LOG, ÷, 2, and ENTER.

: Goto 5
: Disp "NO OF INTERVALS"
: Disp "MUST BE EVEN"
: Goto 4
: Lbl 5
: (B − A)/N→H
: 0→T

This line stores the number zero in the variable T so that we can begin calculating the area for each interval and storing the sum in T.

: B→X
: T + (Y$_1$ − Y$_2$)→T
: A→X

: $T + (Y_1 - Y_2) \rightarrow T$

: $1 \rightarrow I$

: Lbl 1

: $X + H \rightarrow X$

: If IPart$(I/2) = I/2$

: Goto 2

: $T + 4(Y_1 - Y_2) \rightarrow T$

: Goto 3

: Lbl 2

: $T + 2(Y_1 - Y_2) \rightarrow T$

: Lbl 3

: IS $> (I, N - 1)$

: Goto 1

: $TH/3 \rightarrow T$

: Shade$(Y_2, Y_1, 2, A, B)$

The keystrokes for this line are $\boxed{\textbf{2nd}}$, $\boxed{\textbf{PRGM}}$, $\boxed{7}$, $\boxed{\textbf{2nd}}$, $\boxed{\textbf{VARS}}$, $\boxed{2}$, $\boxed{\textbf{ALPHA}}$, $\boxed{\textbf{.}}$, $\boxed{\textbf{2nd}}$, $\boxed{\textbf{VARS}}$, $\boxed{1}$, $\boxed{\textbf{ALPHA}}$, $\boxed{\textbf{.}}$, $\boxed{2}$, $\boxed{\textbf{2nd}}$, $\boxed{\textbf{ALPHA}}$, $\boxed{\textbf{.}}$, $\boxed{\textbf{MATH}}$, $\boxed{\textbf{.}}$, $\boxed{\textbf{MATRIX}}$, $\boxed{\textbf{ALPHA}}$, $\boxed{\textbf{)}}$, and $\boxed{\textbf{ENTER}}$. This instruction shades the area bounded below by the function Y_2, above by Y_1, to the left by A, and to the right by B. The resolution of the shading is 2, so vertical lines are drawn on alternating sets of dots.

: Pause

: Disp "AREA IS"

: Disp T

: End

A Program for Calculating the Area Between Two Curves Using Subroutines

Before executing this program, we must store the functions as Y_1 and Y_2 using the $\boxed{Y =}$ key. The upper function is stored in Y_1. If either of the functions is the x-axis, it is stored as $Y = 0$. The program lets us enter the points of intersection of the functions as the left and right endpoints, A and B, respectively, or it calculates them using a subroutine, a second program that we will enter after completing this one. The program also calculates the area using the Simpson's Rule program as a subroutine. If the curves intersect more than once, the program will only find the area between the first two, left to right, intersection points. In the following program, we will list the keystrokes for and comment on only the instructions that are different from the ones discussed previously.

: AREA
: Disp "DO YOU WANT TO"
: Disp "1 – ENTER A AND B"
: Disp "2 – LET ME FIND"
: Input Q
: If Q = 1
: Goto 3
: Xmin→X
: Prgm6

The keystrokes for this line are $\boxed{\text{PRGM}}$, $\boxed{◁}$, the number or letter of the program, and $\boxed{\text{ENTER}}$. This line transfers execution to the subroutine which finds the intersection point. When the subroutine is finished, execution is transferred back to the next line of the program. Since this is the fifth program we are entering, we are assuming that this subroutine will be stored as the sixth program. If that is not correct, enter the correct letter or number and then press the $\boxed{\text{ENTER}}$ key.

: X→A
: X + .5→X
: Prgm6
: X→B
: Goto 2
: Lbl 3
: Disp "POINT A="
: Input A
: Disp "POINT B="
: Input B
: Lbl 2
: (B+A)/2→X
: If $(Y_1 - Y_2) > 0$
: Goto 1

```
: Disp "YOU ENTERED THE"
: Disp "LOWER FUNCTION"
: Disp "FIRST.  REVERSE"
: Disp "THEM AND RUN"
: Disp "PROGRAM AGAIN."
: End
: Lbl 1
: Prgm4
```

Here we are assuming that Simpson's Rule was entered as the fourth program. If it was not, use the correct number or letter. The program transfers execution to that program to calculate the area and to display the shaded area and the value of the area.

```
: End
```

The program that follows is program six, Prgm6, the subroutine discussed above. It uses Newton's method to find the intersection point.

```
: INTERSEC
: Lbl 1
: If abs (Y₁ − Y₂) < .1
: Goto 2
: X + .01→X
: Goto 1
: Lbl 2
: Y₁ − Y₂→Y
: X − Y/NDeriv(Y₁ − Y₂,.001)→X
: If abs (Y₁ − Y₂) < 1E − 10
: End
: Goto 2
```

Evaluating Improper Integrals

The program AREA, which we stored in the calculator, can be used to approximate the value of an improper integral if we replace ∞ with a large positive number and/or $-\infty$ with a large negative number and use enough intervals. For improper integrals, we would give the calculator the values for A and B. For these types of functions, the graph usually is not very helpful. In each case, the lower function would be $Y_2 = 0$. For example, $\int_{2}^{\infty} \frac{1}{x^3} dx = \frac{1}{8}$. If we let $A = 2$, $B = 200$, and the number of intervals be 100, the area is 0.1416. If we increase the number of intervals to 200, the area is 0.1275. For 300 intervals, the area is 0.1257.

Problem Set 3 ──

1. Find $\int_{2}^{8} x\sqrt{100 - x^2}\, dx$ using Simpson's Rule.

2. Calculate the area between the two functions $f(x) = 5x - x^2$ and $g(x) = x$.

3. Evaluate the improper integral $\int_{1}^{\infty} \frac{dx}{x^2}$

PROBLEMS USING PROGRAMMING APPLICATIONS

Problem Set 4 ──

In Problems $1 - 4$, after the program is written, explain in paragraph form what each line of the program does or means and its interaction with the other lines of the program.

1. Write a program that finds the area bounded by a function and the x-axis using the trapezoidal rule.
2. Write a program that demonstrates the Mean Value Theorem and draws a tangent line to the function at the point $(c, f(c))$.
3. Expand the program for calculating the area between two functions to include functions that intersect at three or more points.
4. Using Problem 49 in Problem Set 10, Chapter 1, write a program that allows you to find the minimum cost of any can if you know the volume and the costs of the top, base and side.

Use the programs given in this chapter or the programs you have written to solve the following problems.

In Problems $5 - 9$, find the roots of the given function to five significant digits.

5. $f(x) = \dfrac{-x^3 - 2x^2 + 3x + 2}{2(x^2 + 1)^{5/2}}$

6. $f(x) = 6\pi x^2 - 200x^{-2}$

7. $f(x) = \left(2x^{-1/2} - \dfrac{3}{x^2}\right)\left(3x^{1/3} - \dfrac{5}{x^2} - 25\right) + \left(4\sqrt{x} + \dfrac{3}{x}\right)\left(x^{-2/3} + \dfrac{15}{x^3}\right)$

8. $f(x) = x \sin x$ for $0 \le x \le 2\pi$

9. $f(x) = x \cos\left(\dfrac{1}{x}\right) - \sin\left(\dfrac{1}{x}\right)$ for $-2\pi \le x \le 2\pi$

In Problems $10 - 14$, draw the tangent line to the function at the given point.

10. $y = x^4 - 32x + 48$ at $(2, 0)$

11. $y = \dfrac{3^x}{7}$ at $\left(5, \dfrac{243}{7}\right)$

12. $y = \sin x + \cos x$ at $\left(\dfrac{\pi}{6}, 1.366025404\right)$

13. $y = \ln x - \log x + \dfrac{1}{x}$ at $(1.1243595, 0.9557036)$

14. $y = \dfrac{1}{3} \ln (\tan 3x + \sec 3x)$ at $(1.918025, -0.1852268)$

Solve Problems $15 - 19$ using numerical integration.

15. $2\pi \displaystyle\int_0^1 (x + x^2 + x^6)\,dx$

16. $\displaystyle\int_2^{10} \dfrac{dx}{x \ln x}$

17. $\displaystyle\int_{\pi/6}^{\pi/3} \dfrac{dx}{\sin x \cos x}$

18. $\displaystyle\int_0^\infty e^{-x^2}\,dx$

19. $\displaystyle\int_0^1 \dfrac{dx}{\sqrt{x}}$

In Problems $20-23$, calculate the area between the two functions.

20. $f(x) = x^3 - 4x$ and $g(x) = x^2 - 4x$; $1 \le x \le 3$

21. $f(x) = \dfrac{1}{x^2 + 3x - 4}$ and $g(x) = (x+3)^2 - 2$ in Quadrant III

22. $f(x) = \cos^5 x$ and $g(x) = \sin^5 x$; $0 \le x \le \frac{\pi}{3}$

23. $f(x) = \dfrac{8}{x} + \dfrac{2}{4x^{1/3}} - e^{-7x}$ and $g(x) = \dfrac{8}{3x^{1/4}} + \dfrac{2}{x} - e^{-10x}$; $1 \le x \le 5$

24. Income per hour for a bookmaker is $30 + 20e^{-0.01t}$. Expenditures are \$34 per hour for t hours. When should the bookie close his operations? What is his profit at that time?

Chapter 3   STATISTICAL CAPABILITIES

The ability to perform statistical and probability calculations is an important part of mathematics. Statistical calculations include such operations as drawing frequency histograms or polygons and calculating the mean, variance, or standard deviation for a given set of data. If the set of data involves two variables, we often need to calculate the correlation coefficient or use a regression model to make projections based on the given data. For continuous probability density functions, we can use the calculator to find the mean, the variance, $P(x > a)$, and $P(x < b)$. All of these calculations can be done using the statistical capabilities of the graphing calculator, so let's start by discussing the various keys we will be using.

USING THE GRAPHING CALCULATOR

The MATH Key

Press the MATH key. The word MATH is highlighted on the first line of the viewing screen. Use the ▷ or ◁ keys to highlight the word PRB.

The PRB Menu

PRB is an abbreviation for the word Probability. Below the first line, the menu is

> 1: Rand
> 2: nPr
> 3: nCr

Rand is an abbreviation for the word Random. Pressing the 1 key followed by the ENTER key causes the calculator to generate a random number between 0 and 1. Since the number is a random number, we have no way of predicting the number that will appear on the viewing screen. This random number can be used in calculations or formulas. For example, if we wish to divide 3 by a random number, the keystrokes are [3], [÷], [MATH], [◁], [1], and [ENTER]. If we perform the operation a second time, we will see that the result is a different number.

nPr is an abbreviation for a permutation of n items r at a time. The keystrokes for calculating a permutation of 8 items 6 at a time are [8], [MATH], [◁], [2], [6], and [ENTER]. The result is 20,160. The formula for performing the calculation is

$$nPr = \frac{n!}{(n-r)!} \text{ with } r \leq n.$$

nCr is an abbreviation for a combination of n items r at a time. The keystrokes for calculating a combination of 8 items 6 at a time are [8], [MATH], [◁], [3], [6], and [ENTER]. The result is 28. The formula for performing the calculation is

$$nCr = \frac{n!}{(n-r)!r!} \text{ with } r \leq n.$$

Problem Set 1 ─────────────────────────────────

1. Find $_{25}P_6$.
2. Find $_{25}C_6$.
3. Explain in paragraph form the difference between nPr and nCr.
4. Read "A Problem Involving Permutations and Combinations" on page 79. Then explain in paragraph form when to use permutations and when to use combinations.

The [STAT] Key

Press the [2nd] key followed by the [MATRIX] key. On the first line of the viewing screen are the words CALC, DRAW, and DATA, with the word CALC highlighted. The five lines that follow define the statistical calculations that we can perform. If we wish to perform one of these calculations, the word CALC must be highlighted. Likewise, whenever the word DRAW is highlighted, the calculator can graph statistical data in three different ways. Before we can perform calculations or graph the data, we must use the DATA function to store the data in the calculator. Use the [▷] or [◁] key to highlight the word DATA.

```
CALC DRAW DATA
1:1-Var
2:LinReg
3:LnReg
4:ExpReg
5:PwrReg
```

The DATA Menu

On the viewing screen we have a four-item menu below the first line. The menu is

1:Edit
2:ClrStat
3:xSort
4:ySort

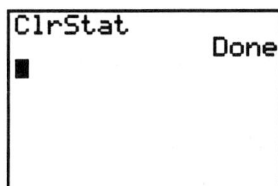

Before we store new data, let's clear any statistical data that might be stored in the calculator. We use the **ClrStat** function to do this. Press the ② key followed by the **ENTER** key. When we press the ② key, the word ClrStat, which is an abbreviation for Clear Statistical Data, appears on the viewing screen. Pressing the **ENTER** key causes the calculator to execute the instruction. The calculator clears all statistical information stored in its memory and the message Done is displayed on the viewing screen. To enter data, press the **2nd** key followed by the **MATRIX** key and use the cursor key to highlight the word DATA.

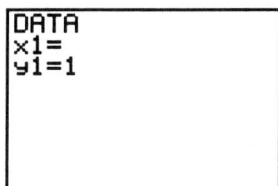

The first function, **Edit**, is used for entering and editing data. Press the ① key. On the first line of the viewing screen is the word DATA. Below this line is a blinking rectangle after $x1 =$, and below that we have $y1 = 1$. We can enter either one-variable or two-variable data. In either the one-variable or two-variable case, the values we store in the calculator are called data points. In the one-variable case, the y values are the frequencies with which the associated x values occur. If we do not enter a y value, the calculator automatically sets the value to be 1. In the two-variable case, the x values are the independent variables and the y values are the dependent variables, or y is a function of x. The calculator interprets the data as one- or two-variable data points when we ask it to perform calculations on the data, not when we enter the data points. We can store up to 150 data points, but the calculator memory is also used to store programs, so this number is decreased by the amount of memory used for storing programs. If we run out of memory, we can free additional memory by erasing one or more programs. Now let's enter some different types of data points to see how it is done.

```
DATA
x1=38
y1=1
x2=37
y2=1
x3=36
y3↓1■
```

First, let's enter 38, 37, 36, 35, 38, 37, 36, 37, 39, and 38 as data points. The blinking rectangle should be after x1=, so enter 38 and press the [ENTER] key twice. For x2, enter 37 and again press the [ENTER] key twice. Repeat this process for each of the numbers. For x10, we should be entering 38, the last number. In each case the calculator automatically sets the entry for y as 1 because we did not enter a value for y. Since three of the numbers − 36, 37 and 38 − repeat, we can take advantage of this by entering the frequency with which a number occurs as the y value. Before we can do this, however, we have to clear out the data we have entered. The keystrokes are [2nd], [MATRIX], [◁], [2], and [ENTER].

```
DATA
x1=38
y1=3
x2=37
y2=3
x3=36
y3↓2
```

Now press the [2nd], [MATRIX], [◁], and [1] keys and enter the data again, noting that 36 occurs twice and 37 and 38 each occur three times. The blinking rectangle is on the line with x1 and the keystrokes are [3], [8], [ENTER], [3], [ENTER], [3], [7], [ENTER], [3], [ENTER], [3], [6], [ENTER], [2], [ENTER], [3], [5], [ENTER], [ENTER], [3], [9], [ENTER], and [ENTER].

```
DATA
x1=50
y1=2
x2=25
y2=4
x3=10
y3↓10
```

The data we entered in each of the examples above are one-variable data, but entering two-variable data is exactly the same as the second example. Let's clear out the data we have stored. The keystrokes are [2nd], [MATRIX], [◁], [2], and [ENTER]. Now, let's enter the points (50, 2), (25, 4), (10, 10), and (5, 19). Press the [2nd], [MATRIX], [◁], and [1] keys. The keystrokes to enter the data points are [5], [0], [ENTER], [2], [ENTER], [2], [5], [ENTER], [4], [ENTER], [1], [0], [ENTER], [1], [0], [ENTER], [5], [ENTER], [1], [9], and [ENTER].

To make corrections on data we have entered, we use the cursor keys to place the blinking rectangle over the number we wish to correct and make the correction. To insert or delete data, use the cursor keys to place the blinking rectangle over the "=" sign of the x or y value where we wish to make the insertion or deletion. If we press the [INS] key, the calculator will move the data points that follow down one pair of values and enter 0 and 1 for the x and y values. We then replace the 0 and 1 with the correct values. If we press the [DEL] key, the calculator erases the values for x and y and moves the points that follow up one pair of values.

```
DATA
x1=5
y1=19
x2=10
y2=10
x3=25
y3↓4
```

Now press the [2nd], [MATRIX], and [◁] keys to get back to the DATA menu. The third function, **xSort**, rearranges the data points so that the x values are in ascending order. Press the [3] key followed by the [ENTER] key. After the word Done appears on the viewing screen, and after pressing the [2nd], [MATRIX], [◁], and [1] keys, we see that the data points are now in the order $(5, 19)$, $(10, 10)$, $(25, 4)$, and $(50, 2)$.

```
DATA
x1=50
y1=2
x2=25
y2=4
x3=10
y3↓10
```

The fourth function in the DATA menu, **ySort**, arranges the data points so that the y values are in ascending order. Pressing the [2nd], [MATRIX], [◁], [4], and [ENTER] keys followed by the [2nd], [MATRIX], [◁], and [1] keys, we see that the data points are now in the order $(50, 2)$, $(25, 4)$, $(10, 10)$, and $(5, 19)$. Now press the [2nd] key followed by the [MATRIX] key and let's examine the CALC menu.

The CALC Menu

The word CALC is an abbreviation for the word Calculation. On the viewing screen we have a five-item menu below the first line. The items are

$$1 : 1 - \text{Var}$$
$$2 : \text{LinReg}$$
$$3 : \text{LnReg}$$
$$4 : \text{ExpReg}$$
$$5 : \text{PwrReg}$$

```
1-Var
x̄=11.28571429
Σx=395
Σx²=8975
Sx=11.52636729
σx=11.36051163
n=35
```

The first item on the list, $1 - \textbf{Var}$, assumes that we have entered one-variable data points into the calculator and calculates the mean and the sample and population standard deviations for the data we have entered. It also gives us the total number of data points we have entered, the sum of the x values, and the sum of the squares of the x values. Press the [1] and [ENTER] keys.

Before discussing the results, we need to remember that the calculator is assuming that the data points are one-variable data. This means that it interprets the values we entered for y as the frequency of the associated values for x. To the calculator, the point $(50, 2)$ means that we have two 50's in our list of data. Likewise, we have four 25's, ten 10's, and nineteen 5's. If we total the values for y, we get 35, which is the value of n in the list above. The total number of data points is $n = 35$. The sum of the x values, $\sum x = 395$, is found by listing all 35 entries for x and finding their sum. The mean, $\bar{x}$, is found by dividing the sum of the x values by the total number of entries, or $\bar{x} = \frac{\sum x}{n} = \frac{395}{35} = 11.28571429$ or $11\frac{2}{7}$. The sum of the squares of the x values, $\sum x^2 = 8975$, is found by squaring each of the 35 entries for x and finding their sum. In statistical analysis, the standard deviation of a set of data can be calculated in two ways. We can discuss the standard deviation of a population or of a sample taken from a population. A population is defined as all of the items we are considering, and a sample is a part of a population. For example, if we are examining the scores of a math class on a particular test, and that class is the only one we are considering, the class is a population. If, however, that class is one of several we are considering, the class is a sample. This distinction is important because the calculation of the standard deviation of a sample, Sx, uses a slightly different formula from that used in calculating the standard deviation of a population, σx. The two formulas are

$$Sx = \sqrt{\frac{n\sum x^2 - (\sum x)^2}{n(n-1)}} \quad \text{and} \quad \sigma x = \sqrt{\frac{n\sum x^2 - (\sum x)^2}{n^2}}.$$

The calculator gives us both answers, $Sx = 11.52636729$ and $\sigma x = 11.36051163$, and leaves it up to us to determine whether our data represent a population or a sample. Before closing this discussion of one-variable statistics, let's make one final observation. The quantities $\sum x$ and $\sum x^2$ are included primarily for information only. They are used to calculate $\bar{x}$, Sx, and σx, but we do not have to perform the calculations. Now press the $\boxed{\text{2nd}}$ key followed by the $\boxed{\textbf{MATRIX}}$ key and let's discuss the remaining items on the list.

Each of the four remaining items on the list, **LinReg**, **LnReg**, **ExpReg**, and **PwrReg**, uses two-variable statistics. In two-variable statistics, each of the data points represents a point on a rectangular coordinate graph, and we are concerned primarily with making predictions based on the data available to us. The expression "making predictions based on the data available to us" means that we would like to be able to determine the relationship that exists between x and y for the data points we have, choose another value for x, and use that relationship to predict the value we would get for y. To be able to do this, we need to find an equation, with y as a function of x, that passes through all the points on our graph. Then, using that equation, we could choose other values for x, substitute them into the equation, and calculate values for y. These y values would be our "predictions" for y. The equation we use for making predictions is called a regression equation. In most cases, however, a regression equation that passes through all the points is impossible to find, so we use the equation that "best fits" the points. This equation usually takes one of four forms. These four forms, listed here, are the four remaining items on the CALC menu.

Abbreviation	Equation Type	Formula
LinReg	Linear	$Y = a + bX$
LnReg	Logarithmic	$Y = a + b \ln(X)$
ExpReg	Exponential	$Y = a(b^X)$
PwrReg	Power	$Y = aX^b$

In each case, the calculator gives us the values of a and b and the value of r, which measures how well the equation "fits" the points. The formulas used for calculating a, b, and r are beyond the scope of this chapter, but we need to discuss the meaning of the value we are given for r. The number r, called the correlation coefficient, can be any number between -1 and 1. In general, the closer r is to -1 or 1, the better the fit. A value for r of -1 or 1 indicates that the equation passes through all the points. The closer r is to zero, the poorer the fit. The sign of r is not important when measuring "goodness of fit." If r is a positive number, the values of y increase as the values of x increase. If r is a negative number, y decreases as x increases. Now let's get an equation of each type for the four data points we have stored in the calculator.

```
LinReg
 a=15.95918367
 b=-.3204081633
 r=-.8483264821
■
```

With CALC highlighted, let's examine the **LinReg** equation. LinReg is an abbreviation for Linear Regression. Press the 2 key followed by the ENTER key. The formulas for a, b, and r are

$$b = \frac{\sum x \sum y - n \sum xy}{(\sum x)^2 - n \sum x^2}, \ a = \frac{\sum y - b \sum x}{n}, \ \text{and}$$

$$r = \frac{n \sum xy - \sum x \sum y}{\sqrt{(n \sum x^2 - (\sum x)^2)(n \sum y^2 - (\sum y)^2)}}.$$

The linear regression equation is $Y = 15.95918367 - .3204081633\,X$. Using this equation, if we choose x to be 15, we can predict a value for y of 11.15306122. The value we are given for r indicates a reasonably good fit, since it is fairly close to -1. We can see the complete regression equation by pressing the VARS key, highlighting the word LR using the cursor keys, and pressing the 4 key. If we have previously stored a value for x in the calculator by entering a number and pressing the STO and X|T keys, pressing the ENTER key will give us the prediction for y.

```
LnReg
 a=28.81740057
 b=-7.268873533
 r=-.9635633765
■
```

Press the 2nd key followed by the MATRIX key to get back to the CALC menu. Let's examine the **LnReg** equation by pressing the 3 key followed by the ENTER key. LnReg is an abbreviation for Logarithmic Regression. This equation uses ln x instead of x to calculate a, b, and r. Thus, in the formula above, $\sum x$ is replaced with $\sum \ln x$, $\sum x^2$ is replaced with $\sum (\ln x)^2$, and $\sum xy$ is replaced with $\sum (\ln x)(y)$. The logarithmic regression equation is $Y = 28.81740057 - 7.268873533 \ln(X)$. Using this equation, if we choose x to be 15, we can predict a value for y of 9.132926137. Furthermore, this prediction for y seems to be better than our prediction using the linear regression equation, because the value for r is closer to -1.

```
ExpReg
 a=18.04613492
 b=.9539256351
 r=-.9604612248
■
```

Press the 2nd key followed by the MATRIX key to get back to the CALC menu. Let's examine the **ExpReg** equation by pressing the 4 key followed by the ENTER key. ExpReg is an abbreviation for Exponential Regression. This equation uses ln y instead of y to calculate a, b, and r. Thus, in the formula above, $\sum y$ is replaced with $\sum \ln y$, $\sum y^2$ is replaced with $\sum (\ln y)^2$, and $\sum xy$ is replaced with $\sum (x)(\ln y)$. The exponential regression equation is $Y = 18.04613492\,(.9539256351^X)$. Using this equation, if we choose x to be 15, we can predict a value for y of 8.89409979. Comparing the values for r, we see that this prediction seems to be better than our prediction using the linear regression equation, but not as good as our prediction using the logarithmic regression equation.

Press the 2nd key followed by the MATRIX key to get back to the CALC menu. Let's examine the **PwrReg** equation by pressing the 5 key followed by the ENTER key. PwrReg is an abbreviation for Power

```
PwrReg
 a=93.62104014
 b=-.9807689606
 r=-.9998582471
■
```

Regression. This equation uses both ln x in place of x and ln y in place of y to calculate a, b, and r. Thus, in the formula above, $\sum x$ is replaced with $\sum \ln x$, $\sum y$ is replaced with $\sum \ln y$, $\sum x^2$ is replaced with $\sum (\ln x)^2$, $\sum y^2$ is replaced with $\sum (\ln y)^2$, and $\sum xy$ is replaced with $\sum (\ln x)(\ln y)$. The power regression equation is $Y = 93.62104014\,X^{-.9807689606}$. Using this equation, if we choose x to be 15, we can predict a value for y of 6.57505909. Comparing the values we are given for r, we see that this prediction seems to be the best of the four, because the value of r is the closest to -1.

Before we leave this section, let's make a few observations. In the logarithmic equation, the values for x must be positive because we are using ln x, and we cannot find the logarithm of zero or a negative number. In the exponential equation, the values for y must be positive because we are using ln y. In the power equation, the values for x and y must be positive because we are using both ln x and ln y. This means that we cannot use the exponential or power equations to predict y if our data include negative values for y. We cannot use the logarithmic or power equations to predict y if our data include negative values for x. If our data include negative values for x and y, we are limited to using only the linear equation to predict y. When we discussed the meaning of the value we are given for r by the calculator, we began the discussion with the words "in general." There are exceptions to the rule that "the closer r is to -1 or 1, the better the fit." One of these occurred in the example above. The best prediction for the value of y is $y = 6.5751$, which we obtained using the power equation. Suppose we are instructed to limit our choice of equations for predicting a value for y to the linear, logarithmic, or exponential equations. Using the linear equation, we had $r = -.8483$ and predicted that $y = 11.1531$. Using the logarithmic equation, we had $r = -.9636$ and predicted that $y = 9.1329$. Using the exponential equation, we had $r = -.9605$ and predicted that $y = 8.8941$. Since the value for r in the logarithmic equation is closer to -1 than in the other two, we would probably choose the logarithmic equation, but when we compare these results with the power equation, which predicted that $y = 6.5751$, we see that the value of y we obtained using the exponential equation is better than the value obtained using the logarithmic equation. This contradicts what we get when we use the rule for comparing r values. Even though it will occasionally give us inaccurate results, the rule "the closer r is to -1 or 1, the better the fit" is still the best guide for choosing which regression equation to use. Now press the $\boxed{\text{2nd}}$ key followed by the $\boxed{\text{MATRIX}}$ key and examine the DRAW menu.

Problem Set 2

1. Given the data $\{(2,2), (4,3), (6,1), (8,4), (10,5), (12,4), (14,6)\}$, find the linear, logarithmic, exponential, and power regression equations.
2. Explain in paragraph form how to determine which equation gives the best fit.

The DRAW Menu

Press the $\boxed{\triangleright}$ key and highlight the word DRAW. Below the first line, the viewing screen reads

> 1:Hist
> 2:Scatter
> 3:xyLine

The DRAW Menu gives us a choice among three ways to display graphically the statistical data we have stored in the calculator. The first of these, **Hist**, which is an abbreviation for Histogram, displays one-variable data as a bar graph. Remember that we have the data points $(50,2)$, $(25,4)$, $(10,10)$, and $(5,19)$ stored in the calculator. The calculator will interpret the second number in each pair as the frequency with which the first occurs. Before we allow the calculator to draw the bar graph, we need to press the $\boxed{\text{RANGE}}$ key and set the variables so that the graph will fit onto the viewing screen. Press the $\boxed{\text{RANGE}}$ key and set **Xmin** $= 0$, **Xmax** $= 60$, **Xscl** $= 5$, **Ymin** $= 0$, **Ymax** $= 20$, and **Yscl** $= 4$. Next we need to press the $\boxed{\text{Y} =}$ key to check for stored equations and erase them with the $\boxed{\text{CLEAR}}$ key. If we do not do this, the graph of the equations will be superimposed on the bar graph. Now press the $\boxed{\text{2nd}}$ and $\boxed{\text{MATRIX}}$ keys, highlight the word DRAW using the cursor keys, and press the $\boxed{1}$ key. The word **Hist** will appear on the viewing screen. Press the $\boxed{\text{ENTER}}$ key to see the histogram. The beginning value and the width of each vertical bar in the histogram are determined by the value of Xscl, so each bar begins with a multiple of 5 and is 5 units wide. The first bar is 19 units tall and starts at $x = 5$. The second bar is 10 units tall and starts at $x = 10$. The third bar is 4 units tall and starts at $x = 25$. The last bar is 2 units tall and starts at $x = 50$. Since the **Hist** function is using one-variable data, we need not enter the data with frequency values. We can enter the 35 values for x in any order and the calculator will divide them into groups using a multiple of the value of Xscl as the beginning value and the value of Xscl as the width of each group, total the number in each group, and then draw the histogram.

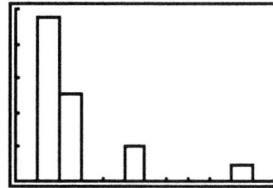

Now press the 2nd key followed by the MATRIX key and highlight the word DRAW using the cursor keys. The second item in the DRAW menu, **Scatter**, uses two-variable data and plots each data point as a coordinate on the viewing screen. In our case it will plot the data points $(50, 2)$, $(25, 4)$, $(10, 10)$, and $(5, 19)$ as four points on the viewing screen. As before, we need to use the RANGE key and check the Range variables to make sure that the data points will fit on the viewing screen. Use the Y = and CLEAR keys to erase any stored functions. To erase the histogram, use the 2nd, PRGM, 1, and ENTER keystrokes. Now press the 2nd and MATRIX keys and highlight the word DRAW using the cursor keys. Press the 2 key and the word **Scatter** will appear on the viewing screen. Press the ENTER key and the four data points will appear on the viewing screen.

Now press the 2nd key followed by the MATRIX key and highlight the word DRAW using the cursor keys. The third item in the DRAW menu, **xyLine**, uses two-variable data and plots the data points like the second item on the list, **Scatter**, but then it connects the data points with a line. Before drawing the graph, we need to erase the previous graph using the keystrokes [2nd], [PRGM], [1], and [ENTER], and we need to arrange the data with the x values in ascending order using the **Xsort** function in the DATA menu. To do this, press the [2nd] and [MATRIX] keys, highlight the word DATA using the cursor keys, and press the [3] key followed by the [ENTER] key. Now, press the [2nd] and [MATRIX] keys, highlight the word DRAW using the cursor keys, and press the [3] key followed by the [ENTER] key. The **xyLine** function can also be used with one-variable data to draw a frequency polygon if we store the frequency with which each data point occurs and sort the data in ascending order.

Problem Set 3

1. For the data {90, 94, 14, 51, 40, 73, 4, 33, 99, 20}, draw
   a. a histogram using your own intervals.
   b. a scatter diagram.
   c. an xy-line.
2. Discuss the relative merits of the histogram, scatter diagram, and xy-line.

The [VARS] Key

Press the [VARS] key. On the first line are the names of five menus that can be accessed by highlighting the name of the menu. These are XY, Σ, LR, DIM, and RNG. Let's discuss the first three menus. We discussed the RNG menu in Chapter 1, and we will discuss the DIM menu in Chapter 4.

The XY Menu

Highlight the word XY. The items in this menu are the values of the statistical results obtained when we execute one of the items in the CALC menu associated with the [STAT] key. The CALC menu was discussed earlier. The values of the items can be displayed or used in expressions. To display the value of an item, press the number key associated with the item and then the $\boxed{\text{ENTER}}$ key. To use the item in an expression, press the number key associated with the item, and the name of the item is copied into the expression. The seven items in this menu are

$$
\begin{aligned}
&1: \text{n} \\
&2: \bar{\text{x}} \\
&3: \text{Sx} \\
&4: \sigma\text{x} \\
&5: \bar{\text{y}} \\
&6: \text{Sy} \\
&7: \sigma\text{y}
\end{aligned}
$$

The first item, **n**, is the number of data points. The second item, $\bar{\text{x}}$, is the mean of the x values. The third item, **Sx**, is the sample standard deviation of the x values. The fourth item, σx, is the population standard deviation of the x values. For one-variable statistical calculations, these items will all be the same as those obtained when we execute the **1-Var** item in the CALC menu. For two-variable statistical calculations, the values of $\bar{\text{x}}$, Sx, and σx will vary. Some regression equations use ln x instead of x. For those using ln x, $\bar{\text{x}}$ is actually the mean of ln x, and Sx and σx are actually the sample and population standard deviations of ln x.

The last three items, $\bar{\text{y}}$, **Sy**, and σy, are the mean, the sample standard deviation, and the population standard deviation, respectively, of the y values. If we are doing one-variable statistical calculations, these items have no meaning. For two-variable statistical calculations, their values will vary because some regression equations use ln y instead of y. For those using ln y, $\bar{\text{y}}$ is actually the mean of ln y, and Sy and σy are actually the sample and population standard deviations of ln y.

The $\sum$ Menu

Press the $\boxed{\text{VARS}}$ key and use the cursor keys to highlight the word $\sum$. The items in this menu are a part of the statistical calculations necessary to execute one of the items in the CALC menu associated with the $\boxed{\text{STAT}}$ key discussed earlier. The values of the items can be displayed or used in expressions. To display the value of an item, press the number key associated with the item and then the ENTER key. To use the item in an expression, press the number key associated with the item, and the name of the item is copied into the expression. The five items in this menu are

$$1: \sum x$$
$$2: \sum x^2$$
$$3: \sum y$$
$$4: \sum y^2$$
$$5: \sum xy$$

The first item, $\sum x$, is the sum of the x values, and the second item, $\sum x^2$, is the sum of the squares of the x values. For one-variable statistical calculations, these items will be the same as those obtained when we execute the **1-Var** item in the CALC menu. For two-variable statistical calculations, the values will vary because some regression equations use ln x instead of x. In those formulas, $\sum x$ is actually $\sum \ln x$ and $\sum x^2$ is actually $\sum (\ln x)^2$.

The third item, $\sum y$, is the sum of the y values; the fourth item, $\sum y^2$, is the sum of the squares of the y values; and the fifth item, $\sum xy$, is the sum of the product of the x values and the y values. If we are doing one-variable statistical calculations, these items have no meaning. For two-variable statistical calculations, their values will vary because some regression equations use ln y instead of y or ln x instead of x. In those cases, $\sum y$ is actually $\sum \ln y$; $\sum y^2$ is actually $\sum (\ln y)^2$; and $\sum xy$ is actually $\sum (\ln x)(y)$, $\sum (x)(\ln y)$, or $\sum (\ln x)(\ln y)$.

The LR Menu

Press the VARS key and use the cursor keys to highlight the word LR. The items in this menu are associated with the regression equation we found using the CALC menu discussed earlier when we examined the STAT key. The values of the first three items can be displayed or used in expressions. To display the value of an item, press the number key associated with the item and then the ENTER key. To use the item in an expression, press the number key associated with the item, and the name of the item is copied into the expression. The four items in this menu are

> 1: a
> 2: b
> 3: r
> 4: RegEQ

The first two items, **a** and **b**, are the same as the values obtained for a and b when we calculated a particular regression equation in the CALC menu. The third item, **r**, is the correlation coefficient for that regression equation. The fourth item, **RegEQ**, is the actual regression equation with the values of a and b inserted into the equation. The equation can be graphed by copying the equation to the $Y =$ list using the $\boxed{Y=}$ key. Assuming there are no equations stored in the $Y =$ list, the keystrokes are $\boxed{Y=}$, $\boxed{VARS}$, $\boxed{\triangleright}$, $\boxed{\triangleright}$, and $\boxed{4}$. We can also evaluate the regression equation for a particular value of x by pressing the $\boxed{STO}$ and $\boxed{X|T}$ keys to store that value in X and then pressing the $\boxed{VARS}$, $\boxed{\triangleright}$, $\boxed{\triangleright}$, $\boxed{4}$, and $\boxed{ENTER}$ keys.

EXAMPLES OF STATISTICAL CAPABILITIES

A Problem Involving Permutations and Combinations

Suppose we have 9 people in a group and we want to do two things: (1) we want to have the group elect a president, vice-president, and secretary and (2) we want to divide the group into committees with three people on each committee. How many different ways can the three officers be chosen, and how many different ways can committees of three people be chosen? To answer the first part of the question we use a permutation of 9 people 3 at a time. The keystrokes are $\boxed{9}$, $\boxed{MATH}$, $\boxed{\triangleleft}$, $\boxed{2}$, $\boxed{3}$, and $\boxed{ENTER}$. The answer is 504. The group can elect the three officers 504 different ways. To answer the second part of the question we use a combination of 9 people 3 at a time. The keystrokes are $\boxed{9}$, $\boxed{MATH}$, $\boxed{\triangleleft}$, $\boxed{3}$, $\boxed{3}$, and $\boxed{ENTER}$. The answer is 84. The group can be divided into committees made up of three people on each committee 84 different ways. This example illustrates the fundamental difference between permutations and combinations. We use a permutation when the order in which we choose objects is important. We use a combination when the order in which we choose objects is not important. Suppose three people in the group are Bill, Sue, and Tom. To elect Tom as president, Sue as vice-president, and Bill as secretary is different from electing Sue as president, Tom as vice-president, and Bill as secretary. In fact, there is a total of six different ways we could elect these people to the offices of president, vice-president, and secretary. In naming officers, order is important, so we use a permutation to find the total number of different ways we can name the three officers. If we name Bill, Sue, and Tom to a committee, the order in which we list the three names is not important, so we use a combination to find the total number of different ways we can name committees made up of three members of the group.

A Problem Involving Histograms, Frequency Polygons, Means, and Standard Deviations

Suppose we have given a standardized test to twelve students and their scores are 38, 37, 36, 40, 35, 40, 38, 37, 36, 37, 39, and 38. Let's enter the scores into the calculator, noting that the scores of 36 and 40 have a frequency of 2 and the scores of 37 and 38 have a frequency of 3. The keystrokes are 2nd, MATRIX, ◁, 2, and ENTER to clear out old data that might be stored in the calculator. To get ready to enter new data, the keystrokes are 2nd, MATRIX, ◁, and 1. The keystrokes for entering the new data are 3, 5, ENTER, 1, ENTER, 3, 6, ENTER, 2, ENTER, 3, 7, ENTER, 3, ENTER, 3, 8, ENTER, 3, ENTER, 3, 9, ENTER, 1, ENTER, 4, 0, ENTER, 2, and ENTER. Let's draw a histogram of the data. Press the Y= key to make sure all functions have been erased. To ensure that the data will appear on the viewing screen in a suitable fashion, set **Xmin** = 34, **Xmax** = 42, **Xscl** = 1, **Ymin** = 0, **Ymax** = 4, and **Yscl** = 1. Now press the 2nd key followed by the MATRIX key, use the ▷ key to highlight DRAW, and press the 1 and ENTER keys to see the histogram of the data. To see a frequency polygon of the data, we must first clear the histogram from the viewing screen. The keystrokes to do this are 2nd, PRGM, 1, and ENTER. The keystrokes for constructing a frequency polygon are 2nd, MATRIX, ▷, 3, and ENTER. If we wish to make the polygon extend all the way to the x-axis on both ends, we can insert the point 34 with a frequency of 0 at the beginning of the data list and add the point 41 with a frequency of 0 at the end of the data list. Then redraw the frequency polygon.

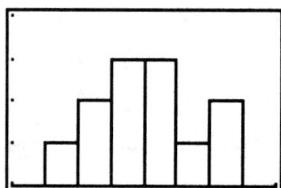

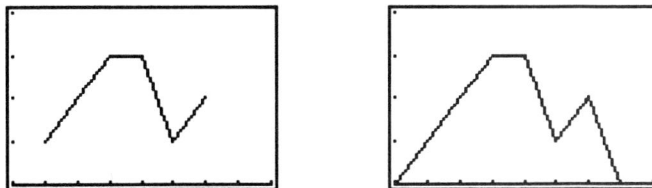

If we wish to find the mean and sample standard deviation of the data, the keystrokes are 2nd, MATRIX, 1, and ENTER. The mean is $\bar{x} = 37.58333333$. The sample standard deviation is Sx = 1.56427929. Another statistic that is used, the variance of the data, is the square of the standard deviation. The sample variance is $Sx^2 = 1.56427929^2 = 2.446969697$. Another way to calculate the sample variance is to press the VARS key followed by the 3 key. This puts Sx on the viewing screen. Next press the x^2 key followed by the ENTER key. The result is the value of the sample variance, 2.446969697.

A Linear Regression Problem

Suppose we know that the marginal cost for certain levels of production of a product is the following, and we are to find the best linear equation for predicting marginal cost for other levels of production.

Units of production(x)	100	200	300	400
Marginal Cost(y)	2400	4100	6000	8500

The best linear equation is the linear regression equation, so we need to store in the calculator the four data points we are given and find the linear regression equation. Press the 2nd key followed by the MATRIX key and highlight the word DATA. Press the 2 key followed by the ENTER key to clear any data we might have stored in the calculator. Press the 2nd key followed by the MATRIX key and highlight the word DATA again. Press the 1 key and enter the data points. The keystrokes are 1, 0, 0, ENTER, 2, 4, 0, 0, ENTER, 2, 0, 0, ENTER, 4, 1, 0, 0, ENTER, 3, 0, 0, ENTER, 6, 0, 0, 0, ENTER, 4, 0, 0, ENTER, 8, 5, 0, 0, and ENTER. Now press the 2nd key followed by the MATRIX key to get the CALC menu. With CALC highlighted, press the $\boxed{2}$ key followed by the $\boxed{\text{ENTER}}$ key. On the viewing screen will be a = 200, b = 20.2, and r = .9959080109.   So the linear regression equation is Y = 200 + 20.2X. The r value of 0.9959 indicates that we have a very good fit for the points.

CALCULATING CONTINUOUS PROBABILITY DENSITY FUNCTIONS

Types of Probability Density Functions

In applied calculus we encounter the integration of certain types of functions which are called continuous probability density functions. These functions will have an area of one square unit over their domain. A simple example of this type of function is the piecewise function

$$y = \begin{cases} \dfrac{1}{180} & \text{for } 0 \le x \le 180 \\ 0 & \text{elsewhere} \end{cases}$$

The graph is a rectangle 180 units long and 1/180 units tall, so its area is 1 square unit. If we calculate the area using an integral, we will have $\int_{0}^{180} \frac{1}{180} dx = 1$.

We can calculate the mean (μ) or expected value $(E(x))$, the variance (σ^2), and $P(x > a)$, $P(x < b)$, or $P(a < x < b)$ of the functions using the following formulas with c and d representing the left and right boundaries of the domain of the function, respectively.

$$\mu = E(x) = \int_{c}^{d} x f(x) dx, \quad \sigma^2 = \int_{c}^{d} (x - \mu)^2 f(x) dx, \quad \text{and } P(x > a) = \int_{a}^{d} f(x) dx$$

To approximate these functions, we will use the program AREA discussed in Chapter 2. We will make two changes in the program SIMPRULE, which is used as a subroutine for the program AREA. Since the graph is not very helpful, we will delete two lines from the program SIMPRULE. These are

$$: \text{Shade}(Y_2, Y_1, 2, A, B)$$
$$: \text{Pause}$$

The keystrokes to get the program into the EDIT mode are $\boxed{\text{PRGM}}$, $\boxed{\triangleright}$, and $\boxed{4}$. This last keystroke assumes we stored the program as Prgm4.

Press the $\boxed{\nabla}$ key until the blinking rectangle is on the first line we wish to delete. The keystrokes to delete the two lines are $\boxed{\text{CLEAR}}$, $\boxed{\text{DEL}}$, $\boxed{\text{CLEAR}}$, and $\boxed{\text{DEL}}$. We can exit the program by pressing the $\boxed{\text{2nd}}$ key followed by the $\boxed{\text{CLEAR}}$ key.

Press the $\boxed{\text{Y}=}$ key. If we are calculating the mean, we will enter $Y_1 = X(f(X))$ where $f(X)$ is the probability density function. For the variance, we will enter $Y_1 = (X - \mu)^2(f(X))$. For calculating a probability, we will enter $Y_1 = f(X)$. In each case, the lower function will be $Y_2 = 0$.

An Example of a Continuous Probability Density Function

The following example is taken from Example 3, Section 10.6, <u>Introduction</u> <u>to</u> <u>Mathematical</u> <u>Analysis</u>, Third Edition, Andre' L. Yandl, Brooks/Cole Publishing Company, 1991.

The Milo Electric Company is manufacturing pocket calculators. The quality control division randomly selects samples and uses these samples continuously for a maximum of 125 hours. If a calculator breaks down within that time, its lifetime (x hours) is recorded. After testing many calculators, quality control has found that the probability density function is given by

$$f(x) = \begin{cases} \frac{4}{1625}(7 - \sqrt[3]{x}) & \text{for } 0 \leq x \leq 125 \\ 0 & \text{elsewhere} \end{cases}$$

First, let's verify that it is a probability density function by approximating the area. We enter the functions

$$Y_1 = \frac{4}{1625}(7 - \sqrt[3]{x}) \text{ and } Y_2 = 0$$

Executing the program AREA, we let $A = 0$, $B = 125$, and use 250 intervals. The area is 1.0001, so it is a probability density function.

Now let's calculate the mean, or expected lifetime, of one of their calculators. We will let $A = 0$, $B = 125$, and use 250 intervals. The functions are

$$Y_1 = \frac{4x}{1625}(7 - \sqrt[3]{x}) \text{ and } Y_2 = 0$$

The expected lifetime is 52.1978 hours of usage.

Using the same values for A, B, and the number of intervals, we can calculate the variance using the functions

$$Y_1 = (x - 52.2)^2\frac{4}{1625}(7 - \sqrt[3]{x}) \text{ and } Y_2 = 0$$

The variance is 1282.1647 hours and the standard deviation, the square root of the variance, is 35.8073 hours.

Next let's calculate the probability that a calculator will last between 45 and 75 hours, or $P(45 < x < 75)$. For this, we will let A = 45, B = 75, and use 250 intervals. The functions will be the original equations.

$$Y_1 = \frac{4}{1625}(7 - \sqrt[3]{x}) \text{ and } Y_2 = 0$$

The probability is 0.2285, so about 23% of the calculators will last between 45 and 75 hours.

As a final example, let's calculate the probability that a calculator will last at least 60 hours, or $P(60 < x < 125)$. Using A = 60, B = 125, and 250 intervals, we find that the probability is 0.3998. About 40% of the calculators will last at least 60 hours.

PROBLEMS USING STATISTICAL APPLICATIONS

Problem Set 4

1. You have a class of 30 students. Generate a random number to represent each student. Order the 30 random numbers from smallest to largest. If you were to generate another set of random numbers for these students, would you obtain the same set of random numbers? Write a paragraph explaining your answer.

2. Find $_8C_0$ and $_8P_0$.

3. Find $_8C_1$ and $_8P_1$.

4. Find $_8C_2$ and $_8P_2$.

5. Find $_8C_8$ and $_8P_8$.

6. Find $_{15}C_{10}$, $_{15}C_5$, $_{15}P_{10}$, and $_{15}P_5$. Explain why $_{15}C_{10} = _{15}C_5$, but $_{15}P_{10} \neq _{15}P_5$.

7. Construct a histogram, or bar graph, for the data {4, 7, 5, 9, 6, 7, 9, 6, 11, 8, 13, 10, 7, 6, 9, 9, 1, 2, 3, 6, 6, 5, 4} using the following intervals of width.

    a.   2 units   b.   3 units   c.   Width you think appropriate

8. Using the data in Problem 7, draw a scatter diagram and an xy-line.

9. Construct a histogram, or bar graph, for the following grades:

Boundaries	Frequency
40 − 49	1
50 − 59	8
60 − 69	20
70 − 79	5

10. Using the data in Problem 9, draw a scatter diagram and an xy-line.

11. The following data represent the ages of the members of a city council: {31, 28, 61, 54, 27, 81, 32}. Put the data in rank order, find the mean, and find the standard deviation treating the data as either a sample or a population. Write a paragraph explaining your reasoning for choosing the sample or the population standard deviation.

12. Find the standard deviation for {66, 68, 71, 72, 69, 69, 73, 70, 63, 66, 70, 74, 70, 68, 73, 70}. Write the "best fitting" regression equation for these data and explain why you feel the equation is the best.

13. The grades on the final exam for Sargent's science class are 70, 82, 70, 79, 70, 78, 70, 81, 64, 70, 85, 90, 82, 59, 69, 70, 85, 60, and 62.

    a.   Draw a histogram, or bar graph, showing the frequency of A's, B's, C's, D's, and F's. (Use your school's grading scale.)

    b.   Find the mean, standard deviation of the sample, and standard deviation of the population.

    c.   Find the linear, logarithmic, exponential, and power regression equations for the data. In paragraph form, explain which of these regression equations is best.

14. A manufacturing firm found its profit over the last six years to be

1985	$11,150,000
1986	$13,000,000
1987	$16,500,000
1988	$19,900,000
1989	$24,000,000
1990	$26,150,000

Find the regression equation that best fits the data.   Make projections of the firm's profits for the next five years and determine the equation's estimate for the profits for 1983 and 1984.

The probability that a particular event will occur is the number of ways that event can occur divided by the total number of possible outcomes. For example, if a fair coin is tossed, the probability of showing a head when it lands is 1/2. There is only 1 way we can get a head, and there are 2 ways in which the coin can land. We say $P(H) = 1/2$. Use this information to answer Problems 15 – 18.

15. From a standard deck of 52 cards, draw 5 cards.   What is the probability of drawing 3 aces and 2 kings?
16. When rolling a standard pair of dice, what is the probability of rolling two numbers whose sum is 7? What is the probability of rolling two numbers whose sum is less than 7?
17. You and seven other people are running for three positions on the school board. The top three vote-getters will be elected to the board. What is the probability that you will be elected?
18. Norred, O'Kuma, Tyssen, and Roberts are four professors at Lee College. Norred has 120 students, O'Kuma has 115 students, Tyssen has 100 students, and Roberts has 100 students.   Many of these students are taking classes from more than one of these professors. The combinations of students and professors are as follows:

      Norred and O'Kuma - 42 students

      Norred and Tyssen - 30 students

      Norred and Roberts - 29 students

      O'Kuma and Tyssen - 25 students

      O'Kuma and Roberts - 35 students

      Tyssen and Roberts - 23 students

      Norred, O'Kuma, and Tyssen - 15 students

      Norred, O'Kuma, and Roberts - 22 students

      Norred, Tyssen, and Roberts - 15 students

      O'Kuma, Tyssen, and Roberts - 18 students

      Norred, O'Kuma, Tyssen, and Roberts - 10 students

   The students of Norred, O'Kuma, Tyssen, and Roberts are given the same questionnaire in each class. The instructions are to fill out the questionnaire only if the student has not filled it out previously. Therefore, if the students follow instructions, each student fills out the questionnaire only once. If a questionnaire is selected, what is the probability that the student whose questionnaire is selected

a.   has only one of the professors?

b.   has only two of the professors?

c.   has only three of the professors?

d.   has all four of the professors?

Problems 19 and 20 have been taken from College Mathematics and Calculus, Second Edition, Karl J. Smith, Brooks/Cole Publishing Company, 1991, Chapter 7 Review Exercises.

19. Find $P(R \cap S)$, $P(R \cup D)$, $P(S \cup \overline{S})$, $P(I \cap \overline{S})$, $P(R \mid S)$, $P(S \mid R)$, $P(D \mid \overline{S})$, and $P(\overline{S} \mid D)$ using the following table.

| | Opinion of President's Policies | | |
Party	Satisfied, S	Not Satisfied, $\overline{S}$	Total
Republican, R	225	175	400
Democrat, D	300	200	500
Independent, I	25	75	100
Total	550	450	1,000

20. Suppose a test for hypoglycemia is given. If a person has hypoglycemia, the test will detect it in 90% of the cases; and if the person does not have hypoglycemia, the test will show a positive result 3% of the time. If we assume that 20% of the population taking the test actually have hypoglycemia, what is the probability that a person taking the test and obtaining a positive result actually has hypoglycemia?

21. Show that $f(x) = \frac{1}{x^2}$ is a probability density function over the interval $x \in [1, \infty)$. Find the expected value of the function.

22. Suppose the probability density function for the average life of a person born after 1940 is $f(x) = 0.10e^{-0.10x}$ where x is the life-expectancy in years. Find the probability that the person will live between 100 and 144 years.

23. YABBOM has several employees to train. From psychology of learning research, they find that the probability density function for learning a random task is $f(x) = \dfrac{1}{(x-3)^2}$ for $x \geq 4$ where x is in days. What is the least number of days required to train 90% of the employees?

24. A CPA for TANSTAAFL, Inc. has developed a system which he believes will make him rich. He will play roulette in sets of three spins. In each of the first two spins, he will bet one unit of money on black. The bet for the third spin will be dependent on the outcome of the first two spins as follows:

   a.   If he wins both of the first two spins, he will not bet.

   b.   If he wins only one of the first two spins, he will bet one unit of money on the color opposite the outcome of the second spin.

   c.   If he loses both of the first two spins, he will bet three units of money on black.

   Let S_1, S_2, and S_3 represent the monetary outcomes of the first, second, and third spins. Calculate $E(S_1)$, $E(S_2)$, $E(S_3)$, and $E(S_1 + S_2 + S_3)$.

25. Given the probability density function
$$f(x) = ((2\pi)(0.8^2))^{-1/2} e^{(-1/2)((x-0.8)/0.8)^2}$$
   find the expected value for $x \geq 1$.

Chapter 4   MATRIX CAPABILITIES

A matrix is a rectangular array of numbers and/or variables. In applied calculus we will use matrices to store data generated by calculator programs. Let's start by discussing the various keys we will need to use.

USING THE GRAPHING CALCULATOR

The MATRIX Key

```
MATRIX EDIT
1:RowSwap(
2:Row+(
3:*Row(
4:*Row+(
5:det
6:ᵀ
```

Press the MATRIX key. On the first line of the viewing screen are the words MATRIX and EDIT, with the word MATRIX highlighted. The six lines that follow define operations we can perform on matrices. Whenever we wish to perform one of these operations, the word MATRIX must be highlighted. Before any operations can be performed, however, we must use the EDIT function to define one or more matrices. Use the ▷ key to highlight the word EDIT.

The EDIT Function

```
MATRIX EDIT
1: [A]   6×6
2: [B]   6×6
3: [C]   6×6
```

On the viewing screen we now see three lines of information below the words MATRIX and EDIT. The calculator can store up to three matrices at a time and perform operations on any or all of them. We can choose the matrix we wish to edit by pressing the number in the first column or by using the ▽ or △ keys to highlight the number and then pressing the ENTER key. The second column lists the names of each of the three matrices. These are set by the calculator and cannot be changed. The third column defines the dimension of the matrix currently being stored by the calculator. The number of rows is listed first, followed by the number of columns. Each of the three matrices may have from 1 to 6 rows and from 1 to 6 columns. The number of rows does not have to equal the number of columns. If a matrix has 3 rows and 4 columns, it is said to be a "3 by 4" matrix. In the third column for that matrix we would have "3 × 4."

Let's start by defining matrix A with 2 rows and 3 columns. On the EDIT screen, either press the $\boxed{1}$ key or use the cursor keys to highlight the 1 and press the $\boxed{\text{ENTER}}$ key. In the upper left corner of the viewing screen we should see [A], followed by the current dimension of the matrix with the first number of the dimension, the number of rows, highlighted. Let's choose our matrix to have 2 rows and 3 columns. Press the $\boxed{2}$ key followed by the $\boxed{\text{ENTER}}$ key. The highlight changes to the second number, the number of columns. Press the $\boxed{3}$ key followed by the $\boxed{\text{ENTER}}$ key. Now the highlight should be on the next line immediately after the "=" sign. We are ready to store the entries in matrix A. Since it has 2 rows and 3 columns, it will have 6 entries in all. Each of these is identified by its position in the matrix, with the row of the entry being named first and the column second. Looking at the numbers to the left of the "=" sign on the highlighted line, we see 1,1. This means we are to enter the number that goes in the first row first column of the matrix. Let's enter -3. The keystrokes are $\boxed{(-)}$, $\boxed{3}$, and $\boxed{\text{ENTER}}$. In the second line we see 1,2 which is the entry in the first row second column. Let's enter 2. For the remaining four entries, let's enter 1, 0, -1, and 5. Now press the $\boxed{\text{2nd}}$ key followed by the $\boxed{\text{CLEAR}}$ key. This takes us back to the Home screen. Now press the $\boxed{\text{2nd}}$ key followed by the $\boxed{1}$ key. This causes an [A] to appear in the upper left corner of the screen, because [A] is to the left just above the $\boxed{1}$ key. Now press the $\boxed{\text{ENTER}}$ key. On the viewing screen we see

```
[A]  2x3
1,1=-3
1,2=2
1,3=1
2,1=0
2,2=-1
2,3=5
```

```
[A]
[  -3   2   1]
[  0   -1   5]
```

Now let's store in matrix B the 3×3 matrix

$$\begin{bmatrix} 1 & 1 & 3 \\ -4 & 2 & 1 \\ 3 & -2 & 0 \end{bmatrix}.$$

Press the MATRIX key, highlight the word EDIT, and press the 2 key. [B] should appear in the upper left corner. To set the dimension, press the 3, ENTER, 3, and ENTER keys. Now let's store the entries in the matrix starting with the first row, then the second row, and then the third row. Remember to press the ENTER key between entries. After finishing, press the 2nd and CLEAR keys to get back to the Home screen. Then press the 2nd and 2 keys, followed by the ENTER key, to see what has been stored.

```
[B]
[ 1   1   3]
[ -4  2   1]
[ 3  -2   0]
■
```

If we have made any errors in storing the entries or in setting the dimension, we can correct them by pressing the MATRIX key, highlighting the word EDIT, and pressing the 2 key. This puts us back into the edit mode for matrix B, and we can use the cursor keys to highlight the error and correct it. When we are finished with our corrections, we can press the 2nd and CLEAR keys to return to the Home screen.

The calculator can store a matrix with up to 6 rows and 6 columns. If the entries are decimals or integers with large absolute values, we may not be able to see all of the matrix on the viewing screen. If that happens, we can use the cursor keys to scroll the screen in order to see the remaining entries.

```
[C]
[ 1   0   2 ]
[ 3  -2   1 ]
[ 4   0  -3]
■
```

Now, let's store the matrix $\begin{bmatrix} 1 & 0 & 2 \\ 3 & -2 & 1 \\ 4 & 0 & -3 \end{bmatrix}$ in matrix C. After storing the entries, press the 2nd and 3 keys, followed by the ENTER key.

Problem Set 1 ————————————————————————

You are teaching a friend to input matrices into the calculator. In paragraph form, describe the process for the following matrices. This should include the keystrokes, the appearance of the viewing screen at each step, and the methods used to correct any errors.

1. $\begin{bmatrix} 4 & 0 & 3 \\ 2 & 1 & 4 \\ 8 & 4 & 3 \end{bmatrix}$ 2. $\begin{bmatrix} 1 & -\frac{1}{5} & -1 \\ \frac{1}{4} & \frac{1}{5} & -\frac{1}{2} \\ 2 & 1 & \frac{1}{6} \end{bmatrix}$

The $\boxed{\text{VARS}}$ Key

Press the $\boxed{\text{VARS}}$ key. On the first line are the names of five menus that can be accessed by highlighting the name of the menu. These are XY, Σ, LR, DIM, and RNG. Let's discuss the DIM menu. We discussed the first three in Chapter 3 and the RNG menu in Chapter 1.

The DIM Menu

Press the $\boxed{\text{VARS}}$ key and use the cursor keys to highlight the word DIM. The seven items in this menu are

> 1: Arow
> 2: Acol
> 3: Brow
> 4: Bcol
> 5: Crow
> 6: Ccol
> 7: Dim{x}

The first six items are the current dimensions (number of rows and number of columns) for the three matrices available to us in the calculator.

The seventh item, **Dim{x}**, is the length of the statistical data list. For one-variable statistical calculations, it is the number of items in the list of data. For two-variable statistical calculations, it is the number of data points.

The values for each of these seven items can be displayed by pressing the number key associated with the item and then the $\boxed{\text{ENTER}}$ key, or they can be used in expressions by pressing the number key associated with the item. The name of the item is copied into the expression.

EXAMPLES OF MATRIX CAPABILITIES

As we use calculator programs to examine limits and derivatives of functions, matrices are a useful way to store data that we would like to examine later. The calculator program can store the data as part of the execution of the program. We can have the program recall the data, or we can recall the data after the program is finished. Let's illustrate this with two programs. Since we discussed programming in Chapter 2, we will list the keystrokes for and comment on only the instructions that refer directly to matrices, except for one instruction in the first program.

A Program for Calculating the Limit of a Function

This program calculates the values of a function as x approaches some number a, which we choose. It stores the values for x and the values for the function as x approaches a from the left in matrix A. It stores the values for x and the values for the function as x approaches a from the right in matrix B. It displays the first matrix, pauses to allow us to examine it, and then displays the second matrix. We can also recall the matrices after the program is finished. We must store the function as Y₁ using the $\boxed{\text{Y} =}$ key.

> : LIMTABLE
> : Disp "A = "
> : Input A
> : A→B
> : 5→N
> : 1→I
> : 5→Arow
> : 2→Acol

The keystrokes for the seventh line of the program are $\boxed{5}$, $\boxed{\text{STO}}$, $\boxed{\text{VARS}}$, $\boxed{\triangleright}$, $\boxed{\triangleright}$, $\boxed{\triangleright}$, $\boxed{1}$, and $\boxed{\text{ENTER}}$. The keystrokes for the eighth line of the program are $\boxed{2}$, $\boxed{\text{STO}}$, $\boxed{\text{VARS}}$, $\boxed{\triangleright}$, $\boxed{\triangleright}$, $\boxed{\triangleright}$, $\boxed{2}$, and $\boxed{\text{ENTER}}$. These lines set the dimension for matrix A. It will have five rows and two columns.

> : Lbl 1
> : B − (.1) ∧ (I − 1)→X
> : (IPart(10000X + .5))/10000→C

The IPart part of the instruction is entered by pressing the $\boxed{\text{MATH}}$ key followed by the $\boxed{\triangleright}$ key and the $\boxed{2}$ key.

> : Y₁→Y
> : (IPart(10000Y + .5))/10000→D
> : C→[A] (I,1)
> : D→[A] (I,2)

The keystrokes for the third line are $\boxed{\text{ALPHA}}$, $\boxed{\text{PRGM}}$, $\boxed{\text{STO}}$, $\boxed{\text{2nd}}$, $\boxed{1}$, $\boxed{(}$, $\boxed{\text{ALPHA}}$, $\boxed{x^2}$, $\boxed{\text{ALPHA}}$, $\boxed{.}$, $\boxed{1}$, $\boxed{)}$, and $\boxed{\text{ENTER}}$. The keystrokes for the fourth line are $\boxed{\text{ALPHA}}$, $\boxed{x^{-1}}$, $\boxed{\text{STO}}$, $\boxed{\text{2nd}}$, $\boxed{1}$, $\boxed{(}$, $\boxed{\text{ALPHA}}$, $\boxed{x^2}$, $\boxed{\text{ALPHA}}$, $\boxed{.}$, $\boxed{2}$, $\boxed{)}$, and $\boxed{\text{ENTER}}$. These two lines store the values for x and the function in the Ith row of matrix A.

> : IS > (I,N)
> : Goto 1
> : Disp [A]
> : Pause
> : Disp " "
> : 1→I
> : 5→Brow
> : 2→Bcol
> : Lbl 2
> : B + (.1) ∧ (I − 1)→X
> : (IPart(10000X + .5))/10000→C
> : Y₁→Y
> : (IPart(10000Y + .5))/10000→D
> : C→[B] (I,1)
> : D→[B] (I,2)
> : IS > (I,N)
> : Goto 2

: Disp [B]
: End

A Program for Graphing Secant Lines Passing Through a Point

This program graphs a function and the secant lines passing through point
a, which we choose, as x approaches a from both sides of a. It illustrates
that the slopes of the secant lines approach the slope of the tangent line.
It also stores the slopes of the secant lines in matrix A, with the slopes
from the left in the first column and the slopes from the right in the
second column, and then displays the matrix. We can also recall the
matrix after the program is finished. We must enter the function as Y₁
using the $\boxed{Y=}$ key. Since the program does not include any instructions
that we have not discussed previously, we will list the program without
comments or keystrokes. The program is

: SECANT
: ClrDraw
: DispGraph
: Pause
: Disp "POINT A IS"
: Input A
: A − .2→Xmin
: A + .2→Xmax
: A→X
: Y₁→B
: B − .2→Ymin
: B + .2→Ymax
: 6→I
: 6→Arow
: 2→Acol
: DispGraph
: Lbl 1
: A − .175 + .025(7 − I)→X
: (IPart(((Y₁ − B)/(− .175 + .025(7 − I)))
  ∗1000))/1000→M

```
: M→[A] (7 − I,1)
: A − 1→C
: B − M→D
: A + 1→E
: B + M→F
: Line(C,D,E,F)
: Pause
: DS < (I,1)
: Goto 1
: Pause
: ClrDraw
: 6→I
: DispGraph
: Lbl 2
: A + .175 − .025(7 − I)→X
: (IPart(((Y₁ − B)/(.175 − .025(7 − I)))
  *1000))/1000→M
: M→[A] (7 − I,2)
: A − 1→C
: B − M→D
: A + 1→E
: B + M→F
: Line(C,D,E,F)
: Pause
: DS < (I,1)
: Goto 2
: Disp [A]
: End
```

PROBLEMS USING MATRIX APPLICATIONS

Problem Set 2 ───

In Problems 1 − 5, use the program for calculating the limit of a function.

1. $\lim\limits_{x \to 0} \dfrac{8^x - 8^{-x}}{8^x + 8^{-x}}$

2. $\lim\limits_{x \to \infty} \dfrac{x - 1}{\sqrt{x^2 + 3x + 2}}$

3. $\lim\limits_{x \to 0} \dfrac{1}{(2x-1)^{2x}}$

4. $\lim\limits_{x \to \infty} xe^{-x}$

5. $\lim\limits_{x \to 0} x \sin \frac{1}{x}$

In Problems $6-10$, estimate the slope of the line tangent to the function at the given point by using the program for graphing secant lines.

6. $f(x) = x^3 - 3x^2 + 3x + 1$ at $x = 2$

7. $f(x) = \tan x + \cot x$ at $x = -\frac{\pi}{4}$

8. $f(x) = e^{(-1/4)x}$ at $x = -4$

9. $f(x) = \ln(4 - x^2)$ at $x = 1$

10. $f(x) = \dfrac{x^2 - x - 2}{x - 2}$ at $x = 0$

Chapter 5   PROJECTS

PROJECTS USING THE GRAPHING CALCULATOR

Introduction

In business, a project is a task with a number of subtasks to be accomplished over a period of time. The project ends with a written report. The problems in this chapter are to be considered as such projects. To complete them, you will be required to do library research, to interview professionals, and to think.

To more easily accomplish the project, you are asked to keep a journal. The journal will contain a record of your work, including the dates and the time spent on the project and the steps you followed. This journal will facilitate the construction of your report. A complete, well-written journal might also be used in solving a future project.

The report should contain a statement of the problem, a summary of the given information, your assumptions for information not given, a breakdown of the subtasks, the literature research, the interviews, the data collected, the analysis of the data, and your conclusions. A statement concerning future use of the information should also be given. The report will also contain items unique to the individual project.

Problem Set

1. You have been hired by the Las Vegas Gambling Club to develop a new game that will give the house 10% of the money bet over a given period of time. Develop the game and then prove, using probability and discrete random variables, that the game meets the Las Vegas Gambling Club's criteria.

2. Electric booster stations are being brought on line for Leek Petroleum at Ogden, Utah. The booster stations are Able, Baker, Charlie, and Dog. Able is 3.75 miles due west of Baker. Charlie is 4688 feet due east and 1083 feet due north of Able. Dog is 7286 feet due west and 3075 feet due south of Baker. How much wire and how many poles are needed to connect the stations in the sequence: Able - Charlie - Dog - Baker. Remember to find out the type of wire to use; the allowance for sag caused by wind, ice, and other weather factors; and other potential variables. After solving this problem, set up a program (with subroutines) to find the amount of wire and number of poles needed regardless of where the construction is to take place.

3. In preparing to be an investment counselor, you decide to play the stock market by analyzing the radius of curvature of the graph made by the stock as it rises and/or falls. Write a program to calculate the curvature at any point. Develop a plan using the curvature to determine when to buy, sell, or stay away from the stock. Select ten stocks and use your plan over a three month period. Make changes in the plan as needed.

4. The Cobb-Douglas production formula is used by economists to set production levels based on available units of labor and capital. You must make a decision on where to build a blue jean manufacturing plant. The two locations are Del Rio, Texas, and Detroit, Michigan. Using the Cobb-Douglas production formula, develop a plan to make your decision in terms of marginal productivity of labor and marginal productivity of capital.

5. You have been hired by the XIT Ranch to develop a population model that is exponential in nature and that will allow you to make decisions concerning the ranch's cattle population in terms of rainfall, price of feed, beef prices, and the time of year.

6. Choose an assembly line that produces at least 1000 units per month. Develop a plan to find the number of workers that will maximize the number of units produced per year.

7. Write a program that will maximize profit if the demand functions for products X and Y are $p_1 = a - bx$ and $p_2 = c - dx$ and the joint cost function for the products is $C(x,y) = exy$, where a, b, c, d, and e are real numbers. Using data from a local industry, try your program.

8. As the negotiator for a labor union, you must develop a cost-benefit model. Choose a local industry and use their data to establish the validity of your model.

Answers to Selected Problems

CHAPTER 1

Problem Set 3

1. Several solutions are possible.
   Xmax − Xmin = 9.5 (Example: **Xmin** = − 4.5 and **Xmax** = 5)
2. Several solutions are possible.
   Ymax − Ymin = 6.3 (Example: **Ymin** = − 2 and **Ymax** = 4.3)

Problem Set 4

1.
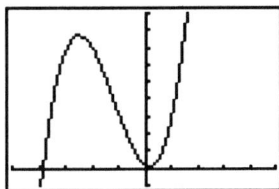
$-10 \leq x \leq 10$
$-10 \leq y \leq 90$

2.

$-10 \leq x \leq 9$
$-20 \leq y \leq 20$

3.
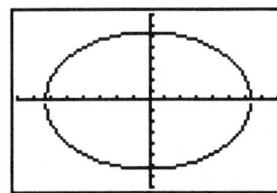
$-40 \leq x \leq 40$
$-40 \leq y \leq 40$

Problem Set 5

1.
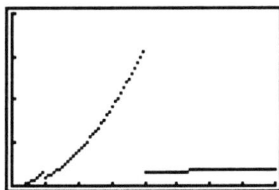
$0 \leq x \leq 8$
$0 \leq y \leq 20$

2.
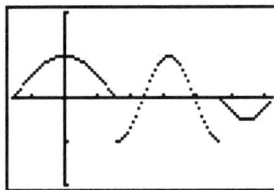
$-\pi/2 \leq x \leq 2\pi$
$-2 \leq y \leq 2$

3.
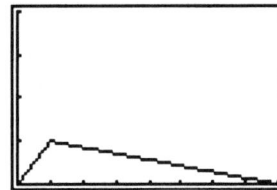
$0 \leq x \leq 1$
$0 \leq y \leq 0.5$

Problem Set 6

4. 2
5. vertical asymptote: $x = -1.2140$
   horizontal asymptote: $y = 1.5$
6. $x = 0, 2$

Problem Set 7

4. relative maximum: $(-1.8165, -3.9113)$; relative minimum: $(-0.1835, -6.0887)$; concave down: $(-\infty, -1)$; concave up: $(-1, \infty)$; inflection point: $(-1, -5)$

5. relative maximum: $(-0.5, 5.0625)$; relative minimum: $(-2, 0)$, $(1, 0)$; concave up: $(-\infty, -1.3660)$; concave down: $(-1.3660, 0.3660)$; concave up: $(0.3660, \infty)$; inflection points: $(-1.3660, 2.2501)$, $(0.3660, 2.2501)$

6. $F(x) = \frac{1}{2}x^2 - x^{-1} + C$

7. $\cos x = 1 - \frac{x^2}{2!} + \frac{x^4}{4!} - \frac{x^6}{6!} + \cdots$

Problem Set 8

1. -37.5

2. 85.5414

3. 10; 1.3953×10^{14}; $69!$

4. a. -81.28 b. -81

   c. -0.2789 d. -82

Problem Set 9

2.

$$0 \le x \le 10$$
$$0 \le y \le 10$$

Problem Set 10

1. $(0,0)$; $(-16.6667, 0)$, $(0, 5)$; $(16.6667, 0)$, $(0, -5)$

2. $(n\pi, 0)$, $(0, 0)$; $(n\pi, 0)$, $(0, 0)$; $(\frac{n\pi}{2}, 0)$, $(0, 0)$

3. $(n\pi, 0)$, $(0, 0)$; $(-\frac{\pi}{2} + n\pi, 0)$, $(0, 0)$; $(\frac{\pi}{2} + n\pi, 0)$, $(0, 0)$

4. $(1, 0)$; $(0.5, 0)$; $(2, 0)$

5. $(0, 0)$; $(0.5671, 0)$; none

6. limit does not exist; discontinuous at $x = 5$

7. 3; discontinuous at $x = 1$

8. 0.6; discontinuous at $x = 2$

9. 0

10. -0.4167; discontinuous at $x = 2$

11. relative maximum: $(-1, 5.6667)$; relative minimum: $(1, 4.3333)$;
    concave down: $(-\infty, 0)$; concave up: $(0, \infty)$; inflection point: $(0, 5)$

12. vertical asymptote: $x = 4$; slant asymptote: $y = x + 2$

13. relative minimum: $(-0.3333, -0.1226)$; concave down: $(-\infty, -0.6667)$;
    concave up: $(-0.6667, \infty)$; inflection point: $(-0.6667, -0.0902)$

14. relative minimum: $(0, 0)$; concave down: $(-\infty, -1.2061)$;
    concave up: $(-1.2061, 1.2061)$; concave down: $(1.2061, \infty)$;
    inflection points: $(-1.2061, 1.5197)$, $(1.2061, 1.5197)$; horizontal asymptote: $y = 0$

15. The equation is not a function. It can be graphed like a function if we solve for y using the
    quadratic formula and enter the results as $Y_1 = \dfrac{-3x + 4\sqrt{25 - x^2}}{25}$ and $Y_2 = \dfrac{-3x - 4\sqrt{25 - x^2}}{25}$. The
    graph is an ellipse. The first function is the upper half of the ellipse. It has a relative maximum
    at $(-3, 1)$ and is concave down over its domain of $(-5, 5)$. The second function is the lower half.
    It has a relative minimum at $(3, -1)$ and is concave up over its domain of $(-5, 5)$.

16. concave up: $(-\infty, 1)$; concave down: $(1, 3)$; concave up: $(3, \infty)$;
    inflection points: $(1, 0.7358)$, $(3, 0.4979)$

17. relative maximum: $(0, 0.6931)$

18. 1   19. 1

20. does not exist   21. 1

22. 0.0156   23. -1.25

24. 4.2403   25. -3.8589

26. $(-10)^3 = -1000$ $(-9)^3 = -729$ $(-8)^3 = -512$
    $(-7)^3 = -343$ $(-6)^3 = -216$ $(-5)^3 = -125$
    $(-4)^3 = -64$ $(-3)^3 = -27$ $(-2)^3 = -8$
    $(-1)^3 = -1$ $0^3 = 0$ $1^3 = 1$
    $2^3 = 8$ $3^3 = 27$ $4^3 = 64$
    $5^3 = 125$ $6^3 = 216$ $7^3 = 343$
    $8^3 = 512$ $9^3 = 729$ $10^3 = 1000$

27. $\sqrt[3]{-10} = -2.15443$ $\sqrt[3]{-9} = -2.08008$ $\sqrt[3]{-8} = -2$
    $\sqrt[3]{-7} = -1.91293$ $\sqrt[3]{-6} = -1.81712$ $\sqrt[3]{-5} = -1.70998$
    $\sqrt[3]{-4} = -1.58740$ $\sqrt[3]{-3} = -1.44225$ $\sqrt[3]{-2} = -1.25992$
    $\sqrt[3]{-1} = -1$ $\sqrt[3]{0} = 0$ $\sqrt[3]{1} = 1$
    $\sqrt[3]{2} = 1.25992$ $\sqrt[3]{3} = 1.44225$ $\sqrt[3]{4} = 1.58740$
    $\sqrt[3]{5} = 1.70998$ $\sqrt[3]{6} = 1.81712$ $\sqrt[3]{7} = 1.91293$
    $\sqrt[3]{8} = 2$ $\sqrt[3]{9} = 2.08008$ $\sqrt[3]{10} = 2.15443$

28. $_{20}P_0 = 1$

$_{20}P_1 = 20$

$_{20}P_2 = 380$

$_{20}P_3 = 6840$

$_{20}P_4 = 116,280$

$_{20}P_5 = 1,860,480$

$_{20}P_6 = 27,907,200$

$_{20}P_7 = 390,700,800$

$_{20}P_8 = 5,079,110,400$

$_{20}P_9 = 6.0949 \times 10^{10}$

$_{20}P_{10} = 6.7044 \times 10^{11}$

$_{20}P_{11} = 6.7044 \times 10^{12}$

$_{20}P_{12} = 6.0340 \times 10^{13}$

$_{20}P_{13} = 4.8272 \times 10^{14}$

$_{20}P_{14} = 3.3790 \times 10^{15}$

$_{20}P_{15} = 2.0274 \times 10^{16}$

$_{20}P_{16} = 1.0137 \times 10^{17}$

$_{20}P_{17} = 4.0548 \times 10^{17}$

$_{20}P_{18} = 1.2165 \times 10^{18}$

$_{20}P_{19} = 2.4329 \times 10^{18}$

$_{20}P_{20} = 2.4329 \times 10^{18}$

$_{20}C_0 = {}_{20}C_{20} = 1$

$_{20}C_5 = {}_{20}C_{15} = 15,504$

$_{20}C_1 = {}_{20}C_{19} = 20$

$_{20}C_6 = {}_{20}C_{14} = 38,760$

$_{20}C_2 = {}_{20}C_{18} = 190$

$_{20}C_7 = {}_{20}C_{13} = 77,520$

$_{20}C_3 = {}_{20}C_{17} = 1140$

$_{20}C_8 = {}_{20}C_{12} = 125,970$

$_{20}C_4 = {}_{20}C_{16} = 4845$

$_{20}C_9 = {}_{20}C_{11} = 167,960$

$$_{20}C_{10} = 184,756$$

29.

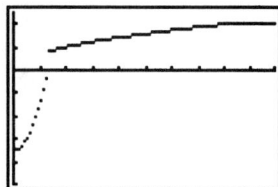

$0 \le x \le 20$
$-10 \le y \le 5$

30.

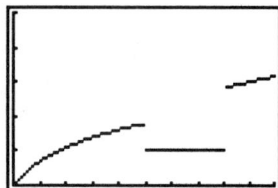

$0 \le x \le 15$
$0 \le y \le 15$

31.

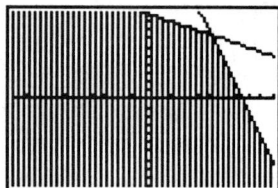

$-3 \le x \le 3$
$0 \le y \le 5$

32.

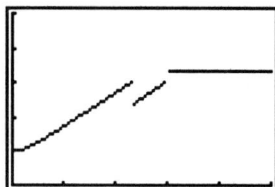

$0 \le x \le 10$
$0 \le y \le 5$

33.

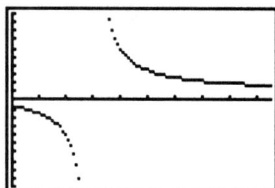

$0 \le x \le 9.5$
$-10 \le y \le 10$

34.

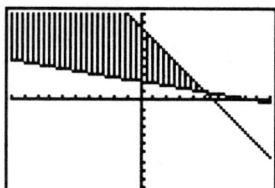

$-20 \le x \le 20$
$-20 \le y \le 20$

35.

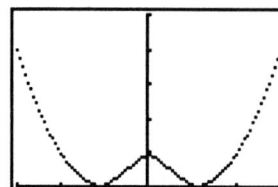

$-10 \le x \le 10$
$-10 \le y \le 10$

36.

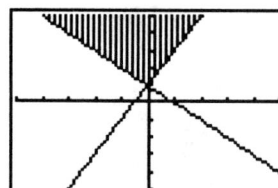

$-5 \le x \le 5$
$-5 \le y \le 5$

37.

38.

39.

$$-10 \le x \le 10$$
$$-10 \le y \le 10$$

$$0 \le x \le 10$$
$$0 \le y \le 10$$

$$0 \le x \le 10$$
$$0 \le y \le 10$$

40. 0.484809620

41. 1.395612425

42. 0.039220713

43. 1.035530314

44. 2.240535650

45. 5 ft 8.84 in × 3 ft 8.84 in × 1 ft 1.58 in

46. $414,138.13

47. 2 days

48. 24 feet

49. $r = 1.9446$ inches; $h = 4.8614$ inches

50. 300 people; $45,000

CHAPTER 2

Problem Set 1

1. The program finds the average of five numbers.

Problem Set 2

1.

2. 8; 13.7097

3.

$$-4.5 \le x \le 5$$
$$-100 \le y \le 100$$

$$-9 \le x \le 10$$
$$-10 \le y \le 10$$

Problem Set 3

1. 241.5347

2. 10.6667

3. With an upper limit of 200, using 250 intervals, the value is 1.0150. Changing the upper limit to 300, and using 500 intervals, the value is 1.0052. The value is approaching 1.

Problem Set 4

5. -2.8136; -0.52932; 1.3429

6. -1.8048; 1.8048

7. 2.1231

8. 0; 3.1416; 6.2832

9. ± 0.05290; ± 0.10493; ± 0.15534; ± 0.29192; ± 1.1623

10.

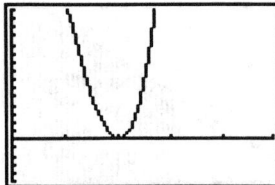

$$0 \le x \le 5$$
$$-5 \le y \le 15$$

11.

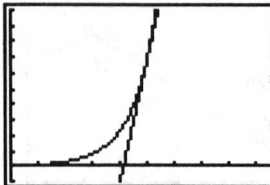

$$0 \le x \le 9.5$$
$$-5 \le y \le 50$$

12.

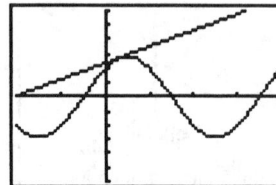

$$-\pi \le x \le 2\pi$$
$$-3 \le y \le 3$$

13.

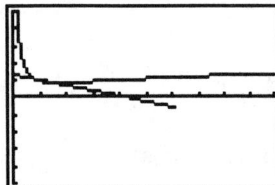

$$0 \le x \le 10$$
$$-5 \le y \le 5$$

14.

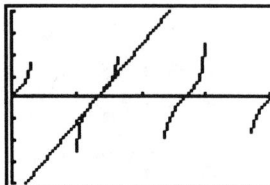

$$0 \le x \le 2\pi$$
$$-2 \le y \le 2$$

15. 6.1336

16. 1.2006

17. 1.0986

18. 0.8862

19. Using 0.1 as a lower limit with 50 intervals, value is 2.0227. Value is approaching 2.

20. 11.3333

21. 2.7045

22. 0.4200

23. 2.7664

24. 6 days 16 hours 56 minutes and 38 seconds; $43,816.68

CHAPTER 3

Problem Set 1

1. 127,512,000

2. 177,100

Problem Set 2

1. Linear: $y = 1 + 0.3214x$
             $r = 0.8082$

    Log: $y = 0.0630 + 1.8359 \ln x$
             $r = 0.7317$

    Exp: $y = (1.4262)(1.1028)^x$
             $r = 0.6878$

    Power: $y = 1.0942x^{0.5484}$
             $r = 0.6109$

Problem Set 4

2. 1 and 1

3. 8 and 8

4. 28 and 56

5. 1 and 40,320

6. 3003; 3003; 10,897,286,400; 360,360

7a.

7b.

8.

9.

10.

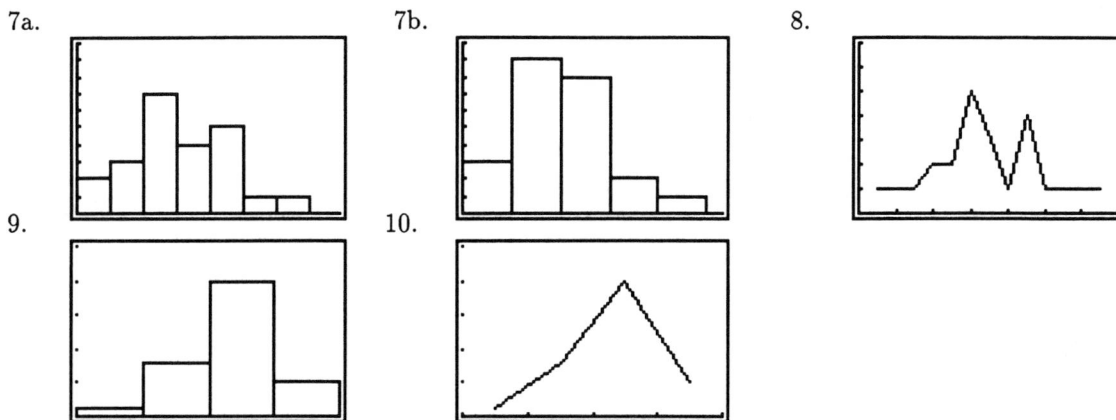

11. $\bar{x} = 44.8571$; $s_x = 20.8601$; $\sigma_x = 19.3127$

12. $s_x = 2.8983$; $\sigma_x = 2.8062$

    Linear: $y = 2.4072 - 0.0090x$ Log: $y = 3.3180 - 0.3632 \ln x$

                $r = -0.0326$ $r = -0.0191$

    Exp: $y = (3.0543)(0.9906)^x$ Power: $y = 13.2661x^{-0.5006}$

                $r = -0.0673$ $r = -0.0523$

13. b. $\bar{x} = 73.4737$; $s_x = 9.1188$; $\sigma_x = 8.8756$

    c.   Linear: $y = 1.3303 + 0.0035x$ Log: $y = -0.2919 + 0.4377 \ln x$

                 $r = 0.0253$ $r = 0.0442$

       Exp: $y = (0.7910)(1.0068)^x$ Power: $y = 0.1191x^{0.5585}$

                 $r = 0.1310$ $r = 0.1481$

14. Linear: $y = -6307478571 + 3182857.143x$ $r = 0.995226$

    Log: $y = -48024518000 + 6325910914 \ln x$ $r = 0.995219$

    Exp: $y = (0)(1.1968)^x$ $r = 0.993531$

    Power: $y = 0x^{357.1229}$ $r = 0.993564$

In the last two equations, when the calculator used the formula to find a, the result to ten decimal places was 0. Thus y in each equation will always be 0, and the equations cannot be used as prediction equations. Using the linear equation, we have

                1991 $29,590,000   1995 $42,321,429

                1992 $32,772,858   1983 $ 4,127,144

                1993 $35,955,715   1984 $ 7,310,001

                1994 $39,138,572

15. $1/108290 = 9.2345 \times 10^{-6}$ 16. 1/6 and 5/12

17. 3/8

18. a. 237/311   b. 34/311
    c. 30/311   d. 10/311

19. 0.225; 0.9; 1; 0.075; 0.4091; 0.5625; 0.4444; 0.4

20. 0.8824   21. $\int_{1}^{\infty}\frac{dx}{x^2} = 1$; Expected value does not exist.

22. 0.0000448   23. 13 days

24. -0.0273; -0.0273; -0.0348; -0.0889

25. 0.6324

CHAPTER 4

Problem Set 2

1. 0   2. 1

3. 1   4. 0

5. 0

6.
$$\begin{bmatrix} 2.572 & 3.472 \\ 2.64 & 3.39 \\ 2.71 & 3.31 \\ 2.78 & 3.23 \\ 2.852 & 3.152 \\ 2.925 & 3.075 \end{bmatrix}$$

slope of tangent is 3

7.
$$\begin{bmatrix} .623 & -.623 \\ .513 & -.513 \\ .406 & -.406 \\ .302 & -.302 \\ .2 & -.2 \\ .1 & -.1 \end{bmatrix}$$

slope of tangent is 0

8.
$$\begin{bmatrix} -.692 & -.666 \\ -.69 & -.669 \\ -.688 & -.671 \\ -.685 & -.673 \\ -.683 & -.675 \\ -.681 & -.677 \end{bmatrix}$$

slope of tangent is $-.6796$

9.
$$\begin{bmatrix} -.589 & -.758 \\ -.601 & -.741 \\ -.614 & -.725 \\ -.626 & -.71 \\ -.639 & -.695 \\ -.652 & -.68 \end{bmatrix}$$

slope of tangent is $-.6667$

10.
$$\begin{bmatrix} 1 & 1 \\ 1 & 1 \\ 1 & 1 \\ 1 & 1 \\ 1 & 1 \\ 1 & 1 \end{bmatrix}$$
slope of tangent is 1